AIChE Centennial: ChE Research and Technology: Past and Future

Topical Conference at the 2008 AIChE Annual Meeting

Philadelphia, Pennsylvania
16 - 21 November 2008

ISBN: 978-1-61567-213-4

Printed from e-media with permission by:

Curran Associates, Inc.
57 Morehouse Lane
Red Hook, NY 12571

Some format issues inherent in the e-media version may also appear in this print version.

Printed by Curran Associates, Inc. (2009)

For permission requests, please contact AIChE
at the address below.

AIChE
3 Park Avenue
New York, NY 10016-5991

Phone: (203) 702-7660
Fax: (203) 775-5177

www.aiche.org

Additional copies of this publication are available from:

Curran Associates, Inc.
57 Morehouse Lane
Red Hook, NY 12571 USA
Phone: 845-758-0400
Fax: 845-758-2634
Email: curran@proceedings.com
Web: www.proceedings.com

AIChE Centennial: ChE Research and Technology: Past and Future

Topical Conference at the 2008 AIChE Annual Meeting

Philadelphia, Pennsylvania
16 - 21 November 2008

ISBN: 978-1-61567-213-4

AIChE

TABLE OF CONTENTS

22a Success and Prospects in Thermodynamics

Peter T. Cummings, Department of Chemical Engineering, Vanderbilt University, Nashville, TN 37235-1604

Text Not Available.

22b Success and Prospects in Transport

Morton Denn, Cuny, Benjamin Levich Institute for Physico-Chemical Hydrodynamics and Department of Chemical Engineering, New York, NY 10031

Text Not Available.

22c Success and Prospects in Catalysis and Reaction Engineering

W. Nicholas Delgass, School of Chemical Engineering, Purdue University, Forney Hall of Chemical Engineering, 480 Stadium Mall Drive, West Lafayette, IN 47907-2100

Text Not Available.

22d Success and Prospects in Process Systems

George Stephanopoulos, Massachusetts Institute of Technology, 77 Massachusetts Avenue, Room 66-440, Cambridge, MA 02139

Process Systems Engineering (PSE) has been at the core of chemical engineering's definition as a discipline since the earliest stages of its inception. Production cost reduction, reduction or elimination of pollutants, and delivery of products at desired specs and amounts, have stimulated an integrated view of processing operations since the latter part of the 19th century. As the notion of unit operations enhanced the intellectual underpinnings for better understanding of the physico-chemical phenomena in a production system, the ability of a system-wide treatment of manufacturing processes started to take off. The revolutionary step in the phenomenal growth of PSE came with the advent of computers. The computer-aided PSE of the second half of the 20th century may be, arguably, credited with the precipitous reduction in unit production cost of all commodity chemicals and materials. At the same time, the expansion of the scope of PSE to include diagnosis, control, and optimization of process operations led to a fast modernization of the chemical processing industries at large, including the pharmaceutical, specialty chemicals, biochemicals, and functional materials industries. Today, PSE has been shaped into a powerful core area of chemical engineering affecting all aspects of CPI including the following: integrated design of products and processes that manufacture them; design of devices for biomedical applications; definition and deployment of sustainable chemicals' production processes; and identification deployment of alternative energy opportunities. As PSE's core influence in CPI has been strengthened and expanded its educational underpinnings have molded generations of chemical engineers, who have taken their skills far beyond the original confines of manufacturing systems with significant impact in many areas of the world economy.

22e Success and Prospects in Bioengineering

Harvey W. Blanch, Chemical Engineering, University of California at Berkeley, 275 Tan Hall, Berkeley, CA 94720

Text Not Available.

The Evolution, Structure and Utility of Separations as an Academic Disciplinary Area

C. Judson King[1]
University of California, Berkeley
MC #4650
Berkeley CA 94720-4650

ABSTRACT

The first great unifying concept of chemical engineering was that of unit operations, put forward by Arthur D. Little and colleagues in the early part of the twentieth century. Many of the unit operations were, in fact, separation processes. With the Manhattan Project, unifying aspects of different methods of separation came to be recognized, and in the decades immediately following World War II were developed and codified so as to create disciplinary underpinnings for separations as a distinct field of knowledge. The history of that transition is traced, and a number of unifying concepts of separations are identified. These concepts enable more efficient and powerful approaches to selection, design, improvement and comparison of separation processes. There are many disciplinary approaches for the pertinent elements of the sciences underlying chemical engineering, but there are few for actual engineering activities. Separations is one such area. Finally, a number of current major separations needs provide vivid examples of why the education of chemical engineers needs to be substantially broadened.

INTRODUCTION

As knowledge has inevitably mushroomed over the years, humankind has sought to organize it and distill general concepts. Organization and generalization make knowledge more powerful and facilitate transmission, understanding and application of knowledge. Codification and development of knowledge on a finer scale have been carried out through disciplines and disciplinary sub-areas. Disciplines have been formed around sets of powerful concepts, generalizations and methods that are applicable to many different situations and which gain both efficiency and strong insights in addressing particular needs and for generating further knowledge.

---

[1] The author is Provost and Senior Vice President – Academic Affairs, Emeritus of the University of California. He is also Professor of Chemical Engineering, Emeritus, Director of the Center for Studies in Higher Education, and Interim Director of the Phoebe A. Hearst Museum of Anthropology of the University of California, Berkeley. E-mail: cjking@berkeley.edu

College degrees in the 19th century tried to cover all of human knowledge, but as knowledge mushroomed there was a move in the early 20th century to combining general education with a specialized major in a discipline. Post-graduate studies then grew, and the idea of general education was largely abandoned in favor of distribution requirements. In engineering, because of the crowded curriculum and the desire to hold the bachelors as the professional degree, general education was largely abandoned, and specialization became almost entirely dominant. I will have more to say about that in the final section of this paper

FROM UNIT OPERATIONS TO SEPARATION PROCESSES[2]

Although chemical engineering as a field arose in the late 1800s[3], the first major unifying concept was that of unit operations, which was put forward by Arthur D. Little and colleagues in an M. I. T. Visiting Committee report in 1915 in the following words.[4] "Any chemical process, on whatever scale conducted, may be resolved into a coordinate series of what may be termed "Unit Operations", as pulverizing, drying, roasting, crystallizing, filtering, evaporating, electrolyzing, and so on. The number of these basic operations is not large and relatively few of them are involved in any particular process. The complexity of chemical engineering results from the variety of conditions as to temperature, pressure, etc., under which the unit operations must be carried out in different processes ..."

Subsequent organizations categorized the unit operations by the phases of matter handled or by whether they dealt with fluid flow, heat transfer, mass transfer or particulate solids. Treybal[5] and Sherwood, et al.[6], focused upon mass transfer and those unit operations involving mass transfer.

Many of the unit operations were, in fact, separation processes, the proportion becoming greater as the field narrowed from all unit operations to those involving mass transfer. Still, the different operations were treated individually in books and courses.

The Manhattan Project of the 1940s, which brought with it the urgent need to separate isotopes so as to produce U-235 and deuterium, provided groundwork for recognition that there are common elements among different separation processes and that there are theoretical underpinnings for separations. The extreme difficulty of these separations led to the development of concepts such as cascade theory, theoretical underpinnings for understanding energy consumption of separations and for gaining energy efficiency, and

---

[2] Another paper by the author treats the development of unit operations and the transition from unit operations to separation processes in more detail: C. J. King, "From Unit Operations to Separation Processes", *Separation and Purification Methods*, 29 (2), 233-245 (2000).
[3] N. A. Peppas, "The First Century of Chemical Engineering", *Chem. Heritage*, 26 (3), 26-29 (2008).
[4] G. G. Brown, *et al.*, "Unit Operations", Wiley, New York, 1950.
[5] R. E. Treybal, "Mass Transfer Operations", McGraw-Hill, nNew York, 1955, 1968, 1980.
[6] T. K. Sherwood, R. L. Pigford & C. R. Wilke, "Mass Transfer", McGraw-Hill, New York, 1975.

distinctions among inherently different classes of separation processes.[7] In a book that saw only limited circulation, Pratt[8] built upon concepts from the Manhattan Project work and extended them to some other separations, notably binary distillation.

An indirect spur toward the recognition of common elements among separation processes was the introduction of the transport phenomena concept with the book by Bird, Stewart and Lightfoot in 1960[9]. This landmark work put fluid flow, heat transfer and mass transfer on a powerful and common fundamental basis. Courses on transport phenomena appeared soon thereafter, with the result that unit operations, and the separation processes that were included within then, stood apart.

As a personal aside, starting in 1963 I was working on my own book, "Separation Processes".[10] Initially, it was to be co-authored with Donald Hanson and to build upon his own class notes on distillation and his work on computer calculations for distillation and extraction.[11] Early on, we addressed the question of whether this should be a book on distillation and possibly some other separation processes, or whether it should be a book on separation processes in general. Without much thought at all, perhaps characteristic of the naivety of a new Assistant Professor, I opted for separation processes in general. Shortly thereafter, Hanson withdrew from the project because of administrative duties and the press of time, and so I went ahead on my own. `As I got further and further into the project over the next several years, I recognized that there are many similarities among different separation processes, along with opportunities for classification and for common means of analysis and synthesis for different separation processes. Consequently, I oriented the book more and more toward separations in general and the concepts that unify them.

I endeavored to establish the common features and functions of separation processes, the factors governing separation factors and selectivity, and the rationales for staging and countercurrent, cross-current, co-current, and fixed-bed designs. Following leads from Hengstebeck[12], I pursued general uses of the McCabe-Thiele (y-x) diagram to illustrate patterns of compositional change and avenues toward improvement of separation processes, including multi-component distillation and non-distillative separation processes. I explored and contrasted which features of various separation processes warranted what analytical, graphical and numerical computation methods. I built forward on the aforementioned work of Cohen, Benedict and Pigford, and Pratt, on energy consumption and treated as relatively new subjects the logic of identifying candidate separation processes and selecting among them, choosing sequences for multi-step

---

[7] K. Cohen, "The Theory of Isotope Separations", National Nuclear Energy Series, Div. III, Vol. 1B, McGraw-Hill, 1951; M. Benedict & T. H. Pigford, "Nuclear Chemical Engineering", McGraw-Hill, New York, 1957.
[8] H. R. C. Pratt, "Countercurrent Separation Processes", Elsevier, 1967.
[9] R. B. Bird, W. E. Stewart & E. N. Lightfoot, "Transport Phenomena", Wiley, New York, 1960.
[10] C. J. King, "Separation Processes", McGraw-Hill, New York, 1971, 1980.
[11] D. N. Hanson, J. H. Duffin & G. F. Somervillee, "Computation of Multistage Separation Processes", Reinhold, New York, 1962.
[12] R. J. Hengstebeck, "Distillation: Principles and Design Procedures", Reinhold, New York, 1961.

separations, and optimizing designs. For the second edition, I dropped the optimization methods and included a chapter on pertinent mass-transfer concepts.

Subsequently, we have seen additional texts and handbooks on separations and separation processes, Engineering Foundation conferences, a National Research Council study, and, significantly, the formation of a Separations Division within the American Institute of Chemical Engineers (AIChE) in 1990 and the reorganization and renaming of Group II of the AIChE Program Committee from Unit Operations to Separations, corresponding to the division that was formed.

UNIFYING CONCEPTS FOR SEPARATION PROCESSES

Here are some of the important unifying features of separation processes.

1. Separation processes can be divided into those that utilize equilibration and those that are rate-based. Equilibration processes can use either energy or mass separating agents.

2. Chemical and/or structured separating agents can be used in the implementations of various mass-separating-agent and rate-based processes. Selection of appropriate chemical agents can be based upon knowledge of solution, complexation, hydrogen-bonding, and solid-state chemistry. Micro- and nano-structures enabling separation can be implemented as holes, pores, cages, chelates, micellar structures and the like, and can be tuned in through control of structural sizes.

3. One can enhance the fundamental degree of separation through counter-currency and staging, generation of reflux, addition of agents at appropriate points, differential migration, and the integration of separation and chemical reaction, such as in distillative esterification processes.

4. The progress of any countercurrent or staged separation can be analyzed by the McCabe-Thiele y-x diagram, which gives direct insights for design improvements.

5. The energy consumption of a separation process is associated with the quantity of separating agent and its ease of regeneration. Energy consumption can be reduced through appropriate concepts of staging, y-x "pinch" alleviation, secondary additions of agents, cascading, appropriate sequencing, and use of the multi-effect concept.

6. Computational methods for different separation processes have common features with the most effective sequencing of calculations and convergence methods depending on physical characteristics of the process.

7. There is logic for selecting among different methods of separation for a particular application.

8. A general understanding of separations facilitates generating entirely new methods of separations. Examples from the past include the development of azeotropic and extractive distillation by Othmer and associates and the development of field-flow methods of separation Giddings and associates, both of which were based upon extensive intuitive insights.

SEPARATIONS AS AN ENGINEERING DISCIPLINARY AREA

As noted earlier, disciplines and sub-disciplines with powerful, unifying concepts make the organization, transmission and application of knowledge more efficient and effective. Chemical engineering has a number of sub-disciples with those characteristics. Examples are transport phenomena, chemical kinetics, thermodynamics, and allied areas such as physical chemistry and organic chemistry. These other sub-disciplines are all sciences, however. They do not convey the essential engineering aspects of the profession.

On the other hand, separations is a sub-discipline that pertains directly to the engineering aspects of chemical engineering. It relates directly to the identification of candidate separation processes for a given need, selection among candidate processes, designing for efficiency and minimum cost, analysis and improvement of process designs, and trouble-shooting of existing processes – all of which are engineering functions. Chemical engineering education tends to emphasize the underlying science over engineering itself. Well designed courses in separation processes teach engineering.

Our profession needs to seek additional sub-disciplinary structures that teach the engineering aspects of the profession. In that way we can boost the engineering end of the curriculum and better prepare our graduates for undertaking engineering functions.

EDUCATING ENGINEERS FOR NOW AND THE FUTURE

Gaining efficiency in the curriculum through inclusion of courses on separations and identification of additional engineering sub-disciplinary areas will help, but it will not do the full job of preparing engineering graduates for their needs and opportunities and rapid changes in today's and tomorrow's worlds. More and more, engineers deal will complex issues and problems that have vital non-engineering and non-scientific dimensions, often with much public interaction and concern. Engineers in the United States will need to be global citizens with awareness of cultures and customs of other countries. Engineers will be well placed to move into management, government, and other, highly multidimensional functions, *if* they can move beyond the elements of science and engineering themselves.

One can recognize these needs even in the more specific context of separations. Think of what will be needed in terms of policies, politics and public education to implement carbon sequestration for ameliorating global warming. Think the same way of the issues of water supply for the population of both the world and parts of the United States, with more and more needs for reclaiming waste and contaminated waters and creating fresh water from sea water. Think of the growing social issues of alleviating contamination of foods, beverages and products of biotechnology. Couple that with the need for cost-efficient and reliable separations associated with the recovery of biologically produced fuels and chemicals from complex and relatively dilute mixtures. Think of the social and policy aspects of controlling worldwide air pollution from many sources beyond just carbon dioxide emissions.

To meet these needs, achieve these opportunities, and lead the most fulfilling lives, engineers in the United States must have a much broader education than is now the case. I have written elsewhere on this subject[13], as have others[14]. Other professions, such as medicine and law, have recognized the value of building upon a general undergraduate education, rather than actually being that undergraduate education. Engineers deserve a similar structure, with breadth in undergraduate education to include the same general-education and distribution requirements as are typical of a liberal arts or College of Letters and Science, education.

These needs are too great to be squeezed into a bachelors-degree education along with the growing body of scientific and engineering knowledge. Engineers should proceed through to a Masters degree for a sufficient professional education, and the Masters should thereby become the accredited professional degree. Engineers even deserve the opportunity to take an undergraduate major that is something other than engineering, as long as they can incorporate sufficient pre-engineering requirements. The "pre-med" concept has served medical education well over the years. It would do the same for engineering, through a more flexible and general bachelors degree that includes the requirement of certain courses specific to preparation for the graduate engineering degree.

It will be argued that industry readily hires engineers at the bachelors level, and so they do in most cases. But this is a place where the interests of the individual engineer diverge from those of corporations. There are many entry-level jobs for which the current bachelors-level engineering education may be sufficient, but the opportunities for engineers to move ahead and have more, and more fulfilling, options open to them are greatly enhanced by a broader and longer education.

Things are happening along these much-needed lines. The civil engineering profession has recognized needs of this sort by defining a Body of Knowledge needed for the profession and launching a move to the Masters degree as the professional and accredited

---

[13] C. J. King, "Let Engineers Go to College", *Issues in Science and Technology*, 22, No. 4, 25-28 (2006). http://www.issues.org/22.4/p_king.html.

[14] J. J. Duderstadt, "Engineering for a Changing World", The Millenium Project, University of Michigan, Ann Arbor MI, December 2007. http://milproj.dc.umich.edu/publications/EngFlex_report/

degree[15]. A recent National Academy of Engineering report[16] recommended that "the baccalaureate degree should be recognized as the "preengineering" degree or bachelor of arts in engineering degree, depending on the course content and reflecting the career aspirations of the student" and that "ABET should allow accreditation of engineering programs of the same name at the baccalaureate and graduate levels in the same department to recognize that education through a "professional" master's degree produces an AME, an accredited "master" engineer". After much debate, ABET has now removed their restriction against accreditation at more than one degree level. This is an important step that enables a transition over time to primary accreditation at the Master's level.

[15] http://www.asce.org/professional/educ/

[16] National Acedemy of Engineering, "Educating the Engineer of 2020: Adapting Engineering Education to the New Century", National Academy Press, Washington DC, 2005.

MEMBRANE SEPARATIONS-100 YEARS OF ACHIEVEMENTS AND CHALLENGES

Tony Fane: UNESCO Centre for Membrane Science & Technology, UNSW, Australia
Singapore Membrane Technology Centre, NTU, Singapore

Over the past 100 years membrane technology has grown from separations techniques of largely theoretical interest to a multibillion dollar industry covering a vast spectrum of applications. Today membrane technology has multiple uses in biomedicine, pharmaceuticals, food and biotechnology industries, the water industry, gas separations including carbon capture, electrochemical applications, analytical and diagnostic uses and other miscellaneous applications. This presentation provides a brief over view of the development of membrane technology and then focuses on the last few decades of membranes in water and wastewater.

In a comprehensive history of membranes up to 1981 Harry Lonsdale[1] estimated at that time the membrane market to be about $1.5bn (corrected to 2008 $) compared with about $ 10m in 1950. Today the market probably exceeds $ 6bn/yr, a 4 fold increase over the past 25 years. The following early history is a précis of the account by Lonsdale and Table 1 is adapted from his text.

Table 1. Early membrane milestones (adapted from [1])

Event	Scientist	Year
Osmosis	Abbe Nollet	1748
Laws of diffusion	Fick	1855
Dialysis, gas permeation	Graham	1861,1866
Osmotic pressure	Traube, van't Hoff	1860-1887
Microporous membranes	Zsigmondy	1907-1918
Distribution law	Donnan	1911
Membrane potential	Teorell, Sievers	1930s
Hemodialysis	Kolff	1944
Gaseous Diffusion	Urey	1940s

It is over 250 years since the Abbe Nollet observed osmosis through a membrane in the form of a pigs bladder. About 100 years later Fick framed his laws of diffusion providing a framework for mass transfer separations. Shortly after this (1860s) Thomas Graham provided us with laws of diffusion in gases and showed early examples of gas separations through rubber; he also demonstrated dialysis. Microporous membranes (MF and UF) were studied by Zsigmondy and Bechold (who introduced the term ultrafiltration) in the early 1900s. Other pioneering developments were demonstrations of 'reverse osmosis' by Manegold and others in the 1920s, Donnans distribution law for charged species and the theory of fixed charge membranes by Teorel et al. in the 1930s. In the early 1940s Kolff demonstrated that dialysis could be used as an artificial kidney. At the same time the gas diffusion process was being developed for UF_6 enrichment). As noted by Lonsdale, by 1950 all the basic principles of membrane transport had been revealed. It is interesting to note that most of this work had been done in Europe, with exception of uranium enrichment, and the pioneers were chemists and physicists. In terms of industrial and environmental separations membranes in the early 1950s were not competitive with alternative processes, such as distillation. The reasons [1] are summarized in Table 2.

Table 2. The non-competitive nature of membrane separations up to 1950-1960.

Non-competitive nature of membranes up to 1950-60
Fluxes were low due to membrane thickness.
Selectivity (separation factor) was relatively poor.
Effective modularization was yet to be developed.
Cheap energy limited benefit of energy efficient membranes.

From about 1960 the modern era of membranes commenced. It is probably no coincidence that from this time Chemical Engineers became significantly involved in membrane process development. It is convenient to discuss recent membrane history as 3 phases, I (1960-1980) the 'beginning of the modern era', II (1980-2000) the 'age of maturation' and III (2000-) the 'age of new challenges'. The invention of the asymmetric integrally skinned cellulose acetate RO desalination membrane by Loeb and Sourirajan [2] in the Engineering School of UCLA was arguably the key event that started academic and commercial interest in membrane separations. Once Loeb and Sourirajan had shown that it was possible to make membranes that were capable of desalting seawater with reasonable fluxes (a 2 order of magnitude improvement on previous membranes) the concept was rapidly taken up by commercial producers. The major applications were in water-scarce regions, such as the Middle East and the Caribbean, and the new RO technology had to compete with more established thermal processes. Significantly the development of a practical RO desalination membrane was a spin off from substantial targeted funding by the US Government through the Office of Saline Water. Towards the end of phase I, a major improvement was made in RO membrane preparation using interfacial polymerization to produce thin film composite (TFC) membranes; John Cadotte (a polymer chemist) patented this in 1997 [3]. The TFC remains the major RO membrane concept today, although it continues to be improved.

The discovery of the L-S RO membrane prompted the development of highly effective UF membranes by chemical engineers, such as Alan Michaels who founded Amicon and also trained many 'membranologists' for the growing industry. It was found that UF membranes could be produced with a wide range of pore sizes, and importantly these membranes could efficiently 'filter' macrosolutes and fine colloids. Two early applications promoted the growth of UF and continue to be major uses today; they are good examples of 'cleaner production'. The first application is the recovery of electrocoat paint colloids in the automotive industry. The evolution of the electrocoat process and UF technology in the 1960s was synergistic. The other important application for UF has been in the dairy industry, with UF used to recover whey proteins, converting a polluting waste into a valuable resource.

The successful development of commercial membrane technology during the 'beginning of the modern era' has relied on many disciplines, but the role of chemical engineers has been central, involving chemical engineers specializing in materials, mass transfer, fluid mechanics, surface phenomena, modeling and process engineering etc. One of the fundamental issues recognized early on by chemical engineers at MIT and elsewhere was the importance of concentration polarization in controlling membrane performance. The need to influence boundary layer conditions close to the membrane required good 'fluid management'. An important example of this was the development of the spiral-wound membrane module (SWM) now the 'work horse' for RO desalination plant and many NF and UF applications. In addition to the SWM the other major concepts are based on cylindrical geometry, either tubes or as hollow fibres (HF). Membranes in HF form have been produced since the 1960s for the range of pressure-driven membrane processes (including RO), as well as for hemodialysis (and other biomedical applications) and gas separations. The largest production until recently has been for hemodialysis.

Hollow fibres are well-suited to gas separations requiring high driving pressures and in this case the feed is applied to the shell with product from the lumen. An important development occurred in the late 1970s when Henis and Tripodi [4] invented a composite HF with a polysulphone (PS) substrate and a thin silicone rubber skin layer; this thin layer effectively plugged imperfections in the PS and produced a fibre of high selectivity and acceptable flux. Gas separation applications include hydrogen recovery (the major use), nitrogen production and CO_2 from methane.

Harry Lonsdale's membrane history [1] ends in 1981 and this covers the 'beginning of the modern era'. It is illuminating to summarize the topics and applications featured in the review (Table 3). The 'future' included Liquid Membranes (coupled and facilitated transport) which has not lived up to expectations. Similarly the elegant immobilized enzyme Membrane Reactors are not in the main stream. There were some notable absentees during this phase I period, and from the anticipated future including membranes applied widely to the water industry, except for RO (and ED) applied to saline waters. However membranes and water has been a major theme in Phase II and III.

Table 3. The status of membranes around 1980 (based on [1])

Process	*Developments*	*Applications*	*Engineering*
MF	Phase inversion membranes Track-etched membranes Stretched membranes	Sterile filtration is 'most important'.	Deadend, but with trend to Cross-Flow filtration, for fluid management.
UF	Hollow fibres, tubular and SWM modules	Pollution control (recover values) Foods, biotech.	Fouling control and fluid management.
Dialysis	Hollow fibre hemodialysis	Disposable artificial kidney.	Mass transfer analysis.
ED	Ion exchange membranes Nafion-type membranes	Desal brackish water. Chlor-alkali in place of Hg cells.	ED Reversal
RO	TFC in SWM Hollow fibres(ar polyamide) SWM & HF in competition.	Desalination. SWM RO about 50% of RO capacity.	Energy recovery devices 'just being introduced'.
Gas Seps	HF compsite (plugged). SWM	H_2, N_2, CO_2	Cascade design

Other Applications	*Comment*
Blood oxygenation	

Membrane electrodes Controlled-release technology	Health-care and agri-food
Future (from 1980 perspective)	
Liquid membranes (couple-facilitated) transport	High selectivity and flux. Pumping 'up hill'.
Membrane reactors (immobilized enzyme type)	Simultaneous conversion and separation.
Pervaporation	Alternative to distillation

The remainder of this brief history will focus on the evolution of membrane technology for water and wastewater in phase II (1980-2000) and on current activities and possible futures in this area (2000 and beyond).

In the period 1960-1980 salty water desalting by RO was the major use of membranes in the water cycle, although it was viewed as a relatively expensive technology best suited to water-scare, energy-rich locations. The prevailing consensus during this period was that membrane technology should only be applied where relatively high value products were involved, such as in the food and biotechnology industries, or in effluent processing to recover values, such as whey proteins or electrocoat colloids. Water was not perceived as a high value product per se.

However in Phase II (1980-2000) a number of factors or drivers have come into play that have significantly changed the perspective. The drivers include tighter regulations on water for consumption, for example based on out breaks of cryptosporidiosis in the early 1990s, and on more regulation of discharge quality. Also we have seen increasing pressures on water resources due to population increase and drought. These challenges have been met by notable developments in membrane technology, so that today membranes play a central role in Water Supply. Sea water desalination, raw water treatment, and water reclamation for reuse all rely on advanced membrane technology to augment our water resources. At this point in time membranes can produce potable water from sea water at US$ 0.5 to 0.7 per m^3, and can produce indirect potable water from municipal effluent for about half this cost, and low pressure (LP) membranes can process raw water at costs similar to conventional treatment [5] with a better quality product. A recent survey of LP membranes shows > 12,000 ML/d installed of which 60% are for water treatment [6]

One of the key developments has been the wide use of the thin film composite membranes in SWM modules for high pressure RO and NF, a trend starting in the late 1980s, and the proliferation of very effective hollow fiber membranes for low pressure UF and MF. There has also been some rationalization of module designs, for example the virtual 'standardisation' of the spiral-wound module (SWM) for high pressure processes. Over the past 10 to 15 years the Submerged Membrane module (in a tank and under suction) has evolved from curiosity to mainstream for low pressure processing and it is of particular interest in water treatment, pretreatment prior to RO and in the membrane bioreactor (MBR). However it has not replaced the contained pressurized hollow fibre module which has some benefits due smaller footprint. Another important paradigm shift has occurred in the application of dead-end operation for membranes processing dilute feeds, such as raw water or settled secondary effluents. This recognizes that energy-demanding crossflow (a feature of 'crossflow-filtration' for fluid management) is not a prerequisite in membrane applications if there are other effective means, such as backflushing with liquid or gas, to periodically remove surface deposits. Water treatment using low pressure UF or MF typically uses hollow fibre membranes with external feed (submerged and some contained) or lumen feed (some contained and pressurized).

The membrane bioreactor (MBR) for waste water treatment was initially developed in Phase I (the late 1960s/ early 70s) but tended to be a 'niche' application due to high pumping energy costs (about 10 kWh/m^3). However the combination of submerged membranes and air sparging has shown that the energy costs can be dramatically reduced to < 1 kWh/m^3 providing the user accepts lower operating fluxes which can be accommodated with lower cost submerged modules (hollow fibres or flat sheet). Major savings have occurred due to improved means of air sparging. Over the past 15 years the implementation of the submerged MBR has been dramatic [7] and the technology continues to grow at >10% per annum. The other major trend in wastewater and membranes has been in reclamation using dual membrane processes, with LP membranes as pretreatment to RO. Major applications can be found in Kuwait (the largest at about 400 Ml/d), Singapore (NeWater) and California (Groundwater Recharge). This approach acknowledges that treated wastewater is a valuable water resource, provided due diligence is applied to its use.

Phase III (2000- and beyond). The challenges facing the future of membrane technology in the water industry relate to product quality, productivity (in terms of fouling and its control), and energy usage and green house gas (GHG) emissions. In the light of these challenges let us look at potential developments in the use of membrane technology in the water cycle, from desalting, water treatment, water reclamation to MBRs.

Desalination. RO is the predominant method of membrane desalting and was a paradigm shift when introduced in the 1960s. The SWM was an early development and has been evolving over the years. The SWM 'Figure of Merit' (FOM) has been defined [8] to illustrate how , over the period 1978 to 2006, improvements in membrane permeability (2.25x) and membrane life (2.3x) and decreases in price per unit area (12x) and salt passage (7x) translate to a 'FOM' increase from 1 to 480. The FOM continues to rise and will be improved by use of Mega modules (16 to 18 inch) and improved thin film composite membranes (see below). Other benefits may come from better spacer design (improved mass transfer with lower pressure drop) as an outcome of CFD analysis (Schwinge et al. [9]). However it seems unlikely that the SWM is the end of the line in module development for SWRO. For several decades RO desalination also used hollow fibres, but this technology is currently sidelined. Nevertheless HF RO could bring intrinsic advantages if revisited, for example with greatly improved (membrane) pretreatment and modules with better fluid management.

Recently there have been significant new developments out of UCLA involving thin film nanocomposite (TFN) RO membranes [10]. These mixed matrix membranes incorporate nanoparticles in the thin polyamide separating film which impart greater hydrophilicity, as well as improving water permeability without loss of retention properties. The TFN concept offers immediate additional routes to improved RO membranes. Longer term 'new generation' RO membranes could come from the carbon nanotubes which promises orders of magnitude increase in permeability [11].

Desalination ~ Energy issues. Significant reductions in energy demand for SWRO have been demonstrated recently in the Affordable Desalination Collaboration project [12]. Using state-of-the-art RO and pressure exchangers the energy has been reduced to as low as 1.58 kWh/m^3, down from the more typical 3 kWh/m^3. The minimum water cost occurs at a recovery of about 50% whereas the minimum energy, and thus minimum GHG impact, occurs at a recovery of about 40%, so GHG minimization comes at a capital cost penalty for SWRO: a paradigm shift will be needed to adopt GHG minimization as the new optimum. There may also be opportunities for further energy reduction if the novel high permeability RO membranes under development are used optimally. Using a 'close to osmotic pressure' feed-side profile it may be possible to save about 35% of the energy. A potential adjunct to SWRO is concentrate processing by Membrane Distillation (MD), but this is only viable in GHG terms if low grade heat (waste or solar) is available. MD has the advantage that flux can be maintained to very high salt concentrations [13] so that overall recoveries > 90% may be feasible. Although MD has been studied for 2 or 3 decades it still awaits a major application, but its potential is attractive. Another membrane process with a long gestation time is Forward Osmosis (FO) which is having a surge of interest [14] as a potentially lower energy approach to desalination. In principle FO could desalinate water at < 1.5 kWh/m^3, provided a suitable membrane can be produced and an easily regenerated draw solute can be identified. Alternatively engineering heuristics suggest that an optimal separation process removes the least abundant species first, so that desalination should involve removal of salt from water rather than vice versa. One membrane process capable of this is electrodialysis, and recent developments are in place to refine this option

Water Treatment. A recent 'membrane' development is the commercial scale application of ceramic MF to water treatment at capacities upto 40 ML/day in Japan. The 15 m^2 modules from NGK have 0.1 micron pore size, and it is claimed that the higher cost of ceramics is offset by much longer lifetime (2 to 5x), greater chemical resistance as well as reportedly higher sustainable fluxes and recovery. For the removal of low concentrations of organics in water treatment the options include NF or low pressure 'hybrid' processes such as MF or UF with adsorbents or photocatalysis. The low pressure hybrid is potentially the lower energy option [15]. In water treatment with low pressure membranes using upstream chemical coagulation the process is typically dead-end with batch cycles and backwash. Cycle time is proportional to $1/(\text{flux})^2$. Each backwash consumes energy and consequently a high flux operation is more energy demanding, although it would require less membrane area. In one comparison conventional economics (minimum costs) suggest a flux in the range 70 to 80 l/m^2hr whereas the minimum energy criteria (minimum GHGs) requires a flux of only 10 to 20 l/m^2hr [15]. In a carbon constrained future there may be a trend to lower fluxes with more investment in membrane area.

Reclamation. Membranes have become the enabling technology for safe and cost effective wastewater reclamation. One of the challenges to RO in this application is biofouling and a future development that may help to combat biofilm development is the thin film nanocomposite (TFN) described earlier [10] but with the incorporation of TiO_2 or similar nanoparticles to reduce biofilms. An alternative approach to reclamation is to incorporate a 'tight' membrane in a membrane bioreactor to achieve higher quality permeate. A radical concept is to use membrane distillation as the separation stage [16].

The MDBR requires the reactor to operate at elevated temperature, such as 50 C, to provide sufficient driving force, so it uses thermophilic bacteria. A limitation of the MDBR process is that to avoid high energy penalty and GHG emissions it must be operated on waste heat. However if waste heat is available the MDBR provides a low GHG approach to reclamation.

Membrane Bioreactors. The MBR is now well established, but improvements may be possible by optimization of bubble size in air sparging. A more radical approach could be to replace or supplement the air scour my mechanical vibrations [17]. In the context of MBRs there are several approaches to 'lower carbon' solutions. The overall wastewater process provides conversion of the biodegradable organic carbon to CO_2 and the conventional aerobic MBR requires net energy input for aeration and membrane fouling control. The incentive to reduce net GHG emissions makes anaerobic processing more attractive if the methane generated can be captured and used to off-set the energy required to run the process. For municipal wastewaters this is more challenging due to the lower carbon load, and potentially lower yield of methane per unit carbon. An interesting option is to combine anaerobic processing (UASB or AnMBR) with polishing in an 'engineered' algal-bacterial MBR that treats the CO_2 and residual carbon. Clearly there are many technical challenges in harnessing the potential energy off-sets in anaerobic processing as well as opportunities for membrane technology. It will be important to quantify the benefits by LCA of the GHG emissions. Finally the direct production of electrical power from biodegradable carbon has been demonstrated in the microbial fuel cell (MFC) [18]. Improved membrane technology will be part of this development.

Decentralization. Decentralized membrane systems, such as water filters and MBRs, can provide benefits from a sustainability perspective [19] as well as potentially tackling the Millenium Development Goals for developing regions. However to apply membranes successfully in decentralized systems we need to make advances in several areas including lower system costs and affordability, minimization of energy demand, maximization of nutrient removal for beneficial use, and developing integrity monitors that are low cost, reliable and remotely accessible.

In Conclusion: the future for membrane separations is very strong. It can be anticipated that membranes will play a key role in a vast range of applications and particularly in critical areas such as the Water Industry (possibly in combination with nanoparticles, engineered biofilms etc), in GHG abatement including CO_2 management and energy (fuel cells etc).

References.

[1] H. K. Lonsdale, The growth of membrane technology, Journal of Membrane Science, 10 (1982) 81-181.

[2] S. Loeb and S. Sourirajan, High flow porous membranes for separating water from saline solutions, US Pat. 3,133,132 (1964).

[3] J. E. Cadotte, Reverse osmosis membrane, US Patent 4,039,440, August 2, (1977).

[4] J. M. S. Henis and M. K. Tripodi, Multicomponent membranes for gas separation, US Patent 4,230,463 (1980).

[5] A. Benedek, Personal Communication, 2008.

[6] D. Furukawa, Survey of LP membranes, National Water Research Institute, 2008.

[7] S. Judd, The MBR Book, Elsevier, 2006.

[8] J. Birkett, and R.Truby. , A figure of merit for appreciating improvements in RO membrane element performance, IDA Newsletter, March 2007.

[9] J. Schwinge, P. R. Neal, D. E. Wiley, D. F. Fletcher and A. G. Fane, Spiral wound modules and spacers: Review and analysis, Journal of Membrane Science, 242 (2004) 129-153.

[10] B.-H. Jeong, E. M. V. Hoek, Y. Yan, A. Subramani, X. Huang, G. Hurwitz, A. K. Ghosh and A. Jawor, Interfacial polymerization of thin film nanocomposites: A new concept for reverse osmosis membranes, Journal of Membrane Science, 294 (2007) 1-7.

[11] J. K. Holt et al, Fast mass transport through sub 2 nanometer carbon nanotubes, Science, 312 (2006) 1034.

[12] R. Truby, Chapter 4, Seawater desalination by ultra low energy reverse osmosis, in N. Li, Fane.A.G., Ho.W.S., and Matsuura,T (Ed.), Advanced Membrane Technology and Applications, John Wiley, New Jersey, 2008.

[13] C. M. Tun, A. G. Fane, J. T. Matheickal and R. Sheikholeslami, Membrane distillation crystallization of concentrated salts--flux and crystal formation, Journal of Membrane Science, 257 (2005) 144-155.

[14] J. R. McCutcheon, R. L. McGinnis and M. Elimelech, A novel ammonia--carbon dioxide forward (direct) osmosis desalination process, Desalination, 174 (2005) 1-11.

[15] A. G. Fane, A. Yeo, A. Law, K. Parameshwaran, F. Wicaksana and V. Chen, Low pressure membrane processes ~ doing more with less energy, Desalination, 185 (2005) 159-165.

[16] J. Phattaranawik, A. G. Fane, A. C. S. Pasquier and W. Bing, A novel membrane bioreactor based on membrane distillation, Desalination, 223 (2008) 386-395.

[17] G. Genkin, T. D. Waite, A. G. Fane and S. Chang, The effect of vibration and coagulant addition on the filtration performance of submerged hollow fibre membranes, Journal of Membrane Science, 281 (2006) 726-734.

[18] B. Logan et al, Microbial Fuel Cells, Environmental Science and Technology (2007).

[19] A. G. Fane and S. Fane, The role of membrane technology in sustainable decentralized wastewater systems, Water Science and Technology, 51 (2005) 317-325.

52c Affinity Separations in the Biotechnology Industry

Chris Lowe, Institute of Biotechnology, University of Cambridge, Tennis Court Road, Cambridge, United Kingdom

Affinity Separatiosn In the Biotecnology Industry

Chemical Engineering and Nuclear Technology at Oak Ridge National Laboratory

During World War II the key Manhattan Project sites were Oak Ridge (Uranium Processing and Pu Processing), Hanford (Pu Production) and Los Alamos (Weapons Development). Oak Ridge was the site for the Gaseous Diffusion Enrichment process, the Electromechanical enrichment process, and the large scale Graphite Reactor and the reprocessing cells that accompanied this in Building 3019. Oak Ridge took the Stagg Field Physics experiment and developed the chemical engineering process that was used as the plan for the Hanford work. After the work Oak Ridge has continued in excellence in Chemical Engineering applications, including operations of several reactors and different flowsheets. Since Oak Ridge is the oldest DOE site working in this area, we also have a strong waste legacy, and have done a lot to treat the nuclear wastes from the early days. Currently ORNL is working with the other labs on the Coupled End to End Chemical Separations activities that will help define the future of nuclear reprocessing and Nuclear Chemical Engineering. This talk will discuss the three activities Oak Ridge was built to address, history of these facilities, how we solved our nuclear waste issues, and the direction Nuclear Chemical Engineering is taking now.

61b Chemical Engineering and Nuclear Technology at Argonne National Laboratory

Ralph Leonard, Chemical Science and Engineering Division, Argonne National Laboratory, 9700 S. Cass Ave, Argonne, IL 60439

Chemical engineering and nuclear technology first met in the Metallurgical Lab and Stagg Field at the University of Chicago at the beginning of World War II. This effort eventually became Argonne National Laboratory. In this work, chemical engineers joined with chemists, physicists, metallurgical engineers, mechanical engineers, mathematicians, and many others to develop nuclear technology. This talk will look back to the beginnings, the problems encountered, the accomplishments, and the plans for the future. Anecdotes will illustrate the application of chemical engineering principles to the development of nuclear technology and the way this nuclear work helped to advance chemical engineering technology as well. Special emphasis will be given to the work in solvent extraction.

This work was supported by the Office of Nuclear Energy of the U.S. Department of Energy under Contract No. DE-AC02-06CH11357 with UChicago Argonne, LLC, Operator of Argonne National Laboratory ("Argonne").

61c a Rich History in Nuclear Materials Processing Technology at Los Alamos National Laboratory

Mark Dinehart, Los Alamos National Laboratory, Los Alamos, NM 87544-3573

Los Alamos National Laboratory has a rich history in Nuclear Materials Processing Technology and chemical engineering has played a key role in the development of this technology. This presentation will review the history of nuclear material processing at Los Alamos and highlight the key contributions chemical engineering has played over the past 50+ years.

Processes and Experiences at the Hanford Site
138546

George Reddick Jr

Introduction

The Hanford site was revealed as a major nuclear site on August 9, 1945 when the atom bomb was dropped on Nagasaki, Japan. Before the announcement in 1945, one of the largest industrial/construction sites in the world had existed in wartime secrecy.

The Manhattan Project chose the Hanford site for plutonium production in January 1943. Construction of two large nuclear reactors and two reprocessing plants began shortly thereafter. The graphite-moderated reactors began operation in 1944. The first plutonium was separated in T Plant in 1945. The plutonium produced at Hanford was used for the first weapon demonstration in Alamogordo, New Mexico in July 1945.

Nuclear materials production continued at Hanford through 1989.

The area chosen for the site encompassed about 700 square miles on both sides of the Columbia River. In 1942, three horticultural communities were located along the Columbia River. White Bluffs and Hanford were two small horticultural communities located near the center of the site and on the west bank of the Columbia River. Hanford gave the site its name. This region west of the Columbia River was known as Hanford before the government began nuclear activities. The town of Richland was near the confluence of the Yakima River and the Columbia River. Richland and Hanford each had a population of about 500. White Bluffs was somewhat smaller. Several other very small communities were located throughout the site on both sides of the river. These communities had survived the depression and had entered into a period of relative prosperity due to the beginning of World War II activities.

Site residents were removed and Richland eventually became the bedroom community for the site.

The site had a history going back about 100 years. George McClellan, of Civil War history, passed through the area as an army surveyor in the 1850s. Ben Snipes, a cattle baron, started running his herds in Eastern Washington between the Canadian border and the Oregon border in the 1850s. The population slowly started to increase in the late 1800s when railroad lines were run through Eastern Washington. Irrigated agriculture began about 1890 in the Yakima valley and about the same time on the Hanford site.

Indigenous populations are assumed to have lived along the river since the last ice age. Glacial ice did not cover the site, but the site was covered by hundreds of feet of water at various times due to the cataclysmic Missoula floods.

Site Selection

The Hanford site satisfied several selection criteria and especially the three principal ones: 1) abundant water for reactor cooling, 2) abundant electrical power to support a large industrial site, and 3) relative isolation for security and safety.

Hanford also had the advantage of relatively low rainfall and a moderate climate. Construction was possible year around with little impact from the weather.

The site selection was made very soon after the decision to place Colonel Leslie Groves in charge of the Manhattan Project. Before his selection, the project had been mostly research and development. He immediately created an emphasis of design/construction and operations. This change was accomplished by hiring major corporations such as duPont, which was placed in charge of Hanford site work.

Nuclear Background

Groves had purchased the site that became Oak Ridge (Tennessee) in September 1942. That site would seek to enrich uranium as weapons material. Seaborg and Wahl had discovered plutonium in February 1941. By July 1941, plutonium was demonstrated superior to uranium as a fissile material. Enrico Fermi had been working with graphite and uranium to create a nuclear chain reaction. In December 1942, Fermi's reactor went critical. In January 1943, the Hanford site was acquired as the location to build reactors and produce plutonium.

Reactors

Nine production reactors were built at Hanford. Three started during World War II. B Reactor started operation in September 1944, which was less than 18 months from the start of design and construction. B Reactor was the world's first full scale reactor.

Three more were added through 1952. The KE and KW Reactors started operation in 1955. N Reactor started in 1963. The production reactors at Hanford were graphite-moderated with horizontal fuel tubes. The fuel used in the production reactors was aluminum-clad uranium metal until N Reactor was started in the early 1960s. The fuel used in N Reactor was zircaloy-clad uranium metal. Fuel was fabricated on site from uranium billets prepared off site.

N Reactor also provided steam for electricity production.

As the first production reactors, all the details associated with operation such as shielding workers from radiation had to be included in the reactor designs. Methods for handling the fuel before and after irradiation were developed. Irradiated fuel was discharged frequently because a relatively large amount of fuel is handled to support plutonium production as contrasted with electrical production.

The fuel was discharged into water basins. The water basins supplied shielding and were convenient for loading casks for shipping to the reprocessing plants. Remote methods and tools were developed for handling the fuel.

Fuel Reprocessing

Hanford used three reprocessing methods. Bismuth phosphate was the initial method. This method consisted of batch precipitation from dissolved fuel using multiple steps. T Plant and B Plant used this method. The plutonium was recovered, purified, and shipped off site. The separated uranium was not purified. The uranium was treated as a waste and transferred to underground storage tanks.

The second method used was the Redox process. This process used methyl isobutyl ketone (hexone) as the solvent in a continuous process. A significant improvement over the Bismuth phosphate process, the Redox process comprehensively recovered the uranium and plutonium.

Due to drawbacks of the Redox process, the AEC sites adopted the Purex process by the mid 1950s. The Hanford Purex plant began operation in early 1956. The Purex process uses continuous solvent extraction to recover and purify uranium and plutonium. The solvent is 30% tri-butyl phosphate in an organic diluent. The Purex process is the standard process that is used throughout the world for reprocessing nuclear fuel.

Uranium Recycling

Uranium recovered from spent fuel reprocessing was recycled for other uses including U^{235} enrichment.

The uranium discharged to the underground storage tanks by the bismuth phosphate process was recovered from the tank waste beginning in 1952. The recovery and purification were accomplished by tri-butyl phosphate solvent extraction.

Waste Management

Opponents' primary objection to the use of nuclear power is the lack of an acceptable waste disposal method. Waste management has always been a significant part (cost) for the fuel cycle. High-level waste is stored in underground tanks at Hanford. With the bismuth phosphate process, the volume of waste per ton of uranium was very large because several batch processes were needed and uranium was not recovered.

The Redox process recovered uranium and the amount of high-level waste per ton of uranium processed was significantly reduced.

The Purex process had high-level waste volumes much less than the previous processes. The Hanford Purex plant was significantly optimized over its operating life to reduce waste volumes and the volume of environmental effluents.

Fission Product Recovery

Various radioisotopes were produced and recovered for selected purposes from the earliest days of the site activities. Tritium, transuranic elements, and other isotopes were produced for defense, medical, and research purposes.

In 1967, fission product recovery took an enormous step change in scale when B Plant was converted to recover cesium and strontium from the high level tank waste. The purpose was to remove Cs^{137} and Sr^{90} from the tank waste to reduce the heat load on the high level waste storage tanks. After the spent fuel is out of the reactor for more than one year, virtually all of the radioactive heat is from cesium and strontium isotopes. After removal of the heat sources, the waste was concentrated to reduce the amount of space needed for waste storage.

The total amount of Cs^{137} and Sr^{90} recovered was well over 100 million curies. The recovered cesium and strontium were purified and double encapsulated for storage in water basins.

Fast Flux Test Facility

The Hanford site also had reactor testing facilities such as the PRTR (Plutonium Recycle Test Reactor). The fuel for this small reactor was plutonium to test the concept of completing the fuel cycle.

The next step was the exploration of the "breeder" reactor concept with the Fast Flux Test Facility, generally known as FFTF. FFTF operated during the 1980s. This reactor used sodium cooling and plutonium fuel. Plutonium-based fuels were successfully tested over a broad range of conditions.

Hanford Today

The size of the Hanford site has shrunk from 700 square miles to less than 600 square miles. Approximately one-half of the site became a national monument in 2000. Part of the national monument is managed by the Washington State Department of Fish and Wildlife. The other part of the monument is managed by the U.S. Fish and Wildlife Service. Most of the rest of the site is undergoing cleanup to remove or stabilize waste sites and to consolidate low level and nonradioactive waste into approved waste sites.

The Department of Energy runs its activities on the site through three major programs: 1) tank waste disposal, 2) site cleanup and waste disposal, and 3) the national laboratory.

In 1965, Hanford activities were diversified among several contractors. The research and development activities became the Pacific Northwest Laboratory. Today the lab is known as Pacific Northwest National Laboratory. Most of the laboratory work in 1965 supported Hanford-related activities including the development of the Fast Flux Test Facility. The laboratory is the largest single employer at the Hanford site. The laboratory activities are significantly diversified and support many programs that are unrelated to the Hanford site.

All of the major operating facilities were shut down by the end of the 1980s. B Plant and Purex were stabilized and deactivated. N Reactor and the other production reactors are going through a process that leads to a removal of all of the reactor facilities and buildings except the reactor core, which is cocooned. The exception is B Reactor, which was recently declared a National Historic Landmark. B Reactor may eventually become a museum as the world's first full-sized reactor.

The major activity on site today is the removal and disposal of waste from the underground storage tanks. The Waste Treatment Plant (WTP) is being built to receive the waste for vitrification. Construction started in about 2001. The various facilities that are part of the WTP complex are expected to start operating within the next decade. Based on the throughput capacity for the WTP, the plant will operate for over 30 years to process and vitrify all of the tank waste.

Other major activities located on the original Hanford site include the Energy Northwest power reactor, Areva's commercial nuclear fuel plant, and LIGO. LIGO is the Laser Interferometer Gravitational-Wave Observatory, which is operated by Caltech and MIT. A duplicate facility is located in Livingston, Louisiana. The purpose of the LIGO is to detect gravity waves.

Hanford Tomorrow

Accurately predicting the future of the Hanford site has not been possible in the last 70 years and will not be possible in the future. There are expectations. Most of the site is expected to be returned to pristine conditions in 50 to 60 years. A portion of the site, particularly the area around the tank farms, will be committed to nuclear waste storage or waste sites. These locations will need to be under administrative control indefinitely for up to 500 years.

Archeological evidence shows the western United States has undergone dramatic climatic changes in the last few thousand years. Prior to historic times, the Columbia River basin has seen some of the most dramatic and cataclysmic floods and geologic events that have been studied. The floods were the result of ice sheet formation and lake formation behind massive ice dams in what is now Montana and British Columbia. New ice sheet formation is not imminent. Fortunately, the recently active volcanoes in the Northwest are not close to the Hanford site.

The forces at work to determine the Hanford site future are similar to those in the battles that are fought daily over drilling versus not drilling for oil and using nuclear energy versus not. The Hanford site has been committed to the pristine versus the industrial as evidenced by the national monument designation. The irony is that if the Hanford site had not been committed to industrial production by the Manhattan Project, the site would be fully developed as a vibrant agricultural and industrial area much as all of the surrounding area is.

The Hanford nuclear activities actually created a false view of what the site was before and what it would have been. The only part of the Columbia River that is not backwater behind a major hydroelectric dam is the so called Hanford Reach along the Hanford site. As another example, a natural gas well existed on the site in the 1930s. Future drilling in a national monument will not be likely. The large elk herd on the site originated from a few stragglers that found the site during a bad winter in the early 1970s. However, elk are not natural to the site or even Eastern Washington. Elk were introduced into Eastern Washington near Yakima in 1913. No elk had lived in Eastern Washington for hundreds of years before that. Removal of the elk herd will not occur even though the herd would not exist if not for the nuclear reservation.

Since part of the site was designated a national monument, the uses for the site are going to be severely limited unless a national emergency occurs. Thus, the Hanford nuclear reservation has created a different paradigm for 600 square miles. The fate of the site is now controlled out of Washington, D.C., by the Washington State congressional delegation, and those intervener and special interest groups that have the ear of those in government.

Insights/Lessons Provided by Hanford

Hanford provided some of the most extraordinary examples of industrial deployment in human history. The initial reactors and reprocessing plants was constructed and started up in less than 18 months. This feat was accomplished even though no similarly-sized nuclear reactors or reprocessing plants had ever operated. The large projects that produced larger reactors and more complex reprocessing plants in the 1950s resulted in operating plants within 5 years.

Today large projects within the government purview require a decade or more to complete even while deploying existing technology. Over such timeframes, government priorities change and political will can be lost. The costs associated with such projects is very large when the inevitable projects changes and overhead costs are applied to the extended project period. Many projects controlled by the Department of Energy during the last several decades were not completed or produced "white elephants" that were not operated or did not operate as intended. The lesson that must be learned is that projects must be needed and must proceed with urgency. Shorter project schedules are more effective (even for complex projects).

Practically, operating plants are optimized after startup. All of the major facilities at Hanford went through significant improvements after startup. Many of the improvements and changes were driven by needs and requirements that occurred after startup. That introduces a philosophy that says that some flexibility needs to be built into the design of nuclear facilities. My experience with newer and ongoing projects is that an effort is made by the projects to provide a perfect facility design with few or no options for making modifications or changes after startup. Not allowing for post project modifications is a perfect way to build an inoperable or ineffective facility.

The amount of laboratory testing, equipment development, and pilot scale testing prior to the deployment of production facilities in the 1940s and 1950s was enormous. Multiple sites and laboratories developed the data and design information needed to support design, construction, and operation. The Hanford site alone had significant facilities for both radioactive and non-radioactive testing through the 1970s. Capabilities from test tube-scale chemistry to full scale equipment testing were available. Processes were demonstrated before deployment. Startups were generally successful and predictable. Laboratory and pilot scale work continued after facility startup to support optimization and modification efforts and new initiatives. Computer modeling is currently in vogue throughout industry. However, for deployment of more complex chemical processes and equipment, thorough development and testing at the appropriate scale worked at Hanford and will still be needed for future deployment of nuclear and chemical facilities.

61e The Role of Chemical Engineering in the Development and Operation of Nuclear Technologies at the Idaho National Laboratory

Jack D. Law, Jack D. Law, and Roger N. Henry. Aqueous Separations and Radiochemistry, Idaho National Laboratory, PO Box 1625, Idaho Falls, ID 83415-3870

Chemical Engineering has played a key role in the development and operation of nuclear fuel cycle technologies at the Idaho National Laboratory over the past 50+ years. The history of these activities will be detailed with a focus on nuclear fuel reprocessing and waste management. A significant effort in the development of nuclear fuel cycle technologies (aqueous and electrochemical separations) is currently in progress as part of the Global Nuclear Energy Partnership and these activities will be detailed. The promising future for chemical engineering at the INL as the nation's nuclear energy laboratory will also be described.

61f Savannah River National Laboratory – a 30 Year History in the Advancement of Environmental Science and Engineering

T. Bond Calloway, Alternative Energy, 999-2W, Savannah River National Laboratory, Savannah River Site, Aiken, SC 29808

The Department of Energy Savannah River Site has excelled at putting (We put to Science to Work™) science to work through the extensive research and development activities conducted by the Savannah River Laboratories and other national laboratories (e.g. Pacific Northwest National Laboratory, Oakridge Ridge National Laboratory, Idaho National Laboratory). Large nuclear waste treatment processes have been developed, designed, constructed and operated with an impressive safety record. This presentation covers the nearly thirty years of process development and project management activities at the Savannah River Site and provides a look into the future activities that are planned for the site.

76a Success and Prospects in Fuels and Energy

F. Emil Jacobs, ExxonMobil Research and Engineering Company, 1545 Route 22 East Room CC332A, Annandale, NJ 08801

Text Not Available.

76b Success and Prospects in Chemicals

Jeffrey J. Siirola, Eastman Research Division, Eastman Chemical Company, PO Box 1972, 200 South Wilcox Drive, Kingsport, TN 37662-5150

Text Not Available.

76c Success and Prospects in Food and Consumer Products

John Leikhim, P&G, X, OH

Text Not Available.

76d Success and Prospects in Pharmaceutical and Health Care Industry

Michael P. Thien, Global Pharmaceutical Commercialization, Merck & Co., Inc., P.O. Box 2000, Rahway, NJ 07065

Text Not Available.

76e Success and Prospects in Advanced Materials and Polymers

Jocelyn E. Scott, DuPont, 1007 Market St., Wilmington, DE 19898

Text Not Available.

Recipes, Models, Chemical Processing and CRE

Octave Levenspiel, Oregon State University, 102 Gleeson Hall, Corvallis, OR 97331-2409

Abstract

The essence of chemical engineering is to come up with recipes and models to help us efficiently make the materials wanted by man. For this purpose, in the first half of the 20th century, studies of reactor behavior focussed on the chemical reaction step and the book by Hougen and Watson was the guiding text for this approach. We call this the American approach.

In the 1950's European engineers broadened their scope to include the heat, mass, diffusion steps that affected the overall rate, plus the flow pattern of the fluid flowing through the reactor. These are all critical for getting a reasonable processing model. We call this the European approach Chemical Reaction Engineering or CRE.

The European approach is more teachable and in general is simpler and more direct, and thus has been widely adopted by engineers working in all sorts of areas.

In the coming century the dominant emphasis in chemical engineering will shift from the processing of petroleum to the processing of some other feed material. I suspect that it will be coal. I will end this talk with thoughts on ways to process coal to syngas.

This paper sketches this story.

Chemical Reaction Engineering: Looking towards the Future from a Molecular Perspective

Alexis T. Bell, Department of Chemical Engineering, UC Berkeley, 373 Tan Hall, Berkeley, CA 94720

Abstract

During the latter half of the twentieth century, the field of chemical reaction engineering made continuous advances in the development of methods for describing the performance of chemical reactors. The development of progressively more efficient and robust methods for solving large sets of partial differential equations and for handling complex reactor geometries and boundary conditions have been essential to achieving these advances. The solution of larger and more complex problems has been enabled, as well, by in advances in computer speed and memory. The past twenty years have also witnessed the development of increasingly accurate and efficient methods for the prediction of adsorption isotherms and diffusion coefficients for molecules in a various micro- and mesoporous materials and for the prediction of the reaction energetics and dynamics from first principles. As a result, it has become possible to think about chemical reaction engineering from both a molecular and a macroscopic perspective. Construction of the bridges between these perspectives has begun to occur, and one can now envision a future in which it will be possible to explore the performance of reactors, beginning with a molecular understanding of the reactants and catalysts together with inputs about the pore structure, connectivity, etc. of the catalyst and the geometry and flow configuration of the reactor. This talk will illustrate some of the advances made in the in the ab initio prediction of reaction kinetics and molecular transport, and will show how such knowledge can be combined to predict catalyst performance at the catalyst pellet level.

Heat management in chemical reactors-Solved and unsolved problems.

Dan Lus, University of Houston,
Houton, TX

Proper management of the heat generated in chemical reactors is essential for the safe operation, start-up and control of any chemical reactor. Lilenroth, a DuPont employee, pointed out already in 1922 that the heat generation in a chemical reactor may lead to multiple steady states, some of which correspond to a runaway. Since that pioneering work, significant advances have been made in gaining an understanding of the impact of start up policy or transient perturbations on the dynamic behavior, stability and sensitivity of chemical reactors. A particular large body of knowledge and understanding of this has been generated by Neal Amundson and his collaborators in Minnesota. Some dynamic features of chemical reactors, such as the wrong-way behavior (a temporal temperature increase caused by a sudden decrease of the feed temperature to a packed bed reactor), are counterintuitive. To circumvent these undesired and unexpected transient temperature rise the design a special has to be applied which incorporate the ability of predicting this complex dynamic response.. The interaction between the temperature change and the fluid properties can have a major impact on the behavior of some reactors and in some cases the reaction rate. One noted example are polymerization reactors in which a change in a temperature induced change in the fluid viscosity affects the ratio between the rate of polymerization initiation, propagation and termination reactions. This in turn, may lead to a drastic undesired change in the properties of the produced polymer. Temperature induced change of the physical properties can lead to thermoflow multiplicity in multi-tube reactors, i.e., cause a nonuniform flow rate in the many catalyst packed tubes of a multi-tube packed bed reactor. This can have a deleterious impact on the yield of the desired product and cause non-uniform deactivation of the catalyst in the various tubes, requiring a early regeneration. The self-ignition of coal piles is another example of the interaction between transport phenomena, physical properties and runaway.

A reliable heat management of a chemical reactor is essential for the safe start, operation and control of a chemical reactor. Most existing rules and insightful guidance about heat management of chemical reactors rules are concerned with continuous-flow reactors. However, many industrial highly exothermic reactions are carried out in batch and semi-batch reactors. There have been many reported cases f reported explosions and runaway in these reactors, including some early ones at the DuPont company during the production of explosives. There is a need to develop similar design rules for the optimal safe feed of reactants to these reactors, i.e, what should be the proper feed rate of the various reactants and the temporal change of the coolant temperature that will lead to safe and optimal production at the desired products.

Proper design of heat management is critical for the reliable operation of autothermal chemical reactors in which the heat generated by the chemical reaction is utilized to preheat the feed to the reactor. It is well known that these reactors ma attain in certain case rather different steady states and it is essential to use a proper start up procedure in order to attain the desired one and apply a proper control to maintain it. In some applications a periodic operation may enhance the energy utilization of the process. For example, a reverse flow operation enables circumventing the need to add fuel during the catalytic combustion of feeds containing a low concentration of pollutants. This reactor may attain different periodic states and it is important to devise the proper start up and operation procedure. In some cases this reactor attains a complex periodic motion, such as a preriod-n or even chaotic operation, rather than the simple period-1 operation. A remaining important unsolved challenging problem is how to use reverse-flow operation to conduct an endothermic reaction using the heat generated by an exothermic one.

The local overheating may lead to major operation problems in industrial reactors. There is an important need to enhance our understanding and ability to predict and to circumvent the formation of local hot zones in a catalytic reactors, in which the overall heat management predicts safe operation. For example, it has been reported that local overheating of single catalytic pellet may occur in fluidized bed reactors during the polymerization of polyolefins. This overheating may lead to local melting of the growing polymer particle and subsequent formation of polymer "sheets", requiring shut-down of the reactor. It is still not established what exactly causes this undesirable events, a knowledge needed in order to circumvent their occurrence. It is also well established that local hot spots ay form in large diameter packed bed reactors under conditions that the average temperature is closely controlled. When these hot zones form in the center of the reactor they cause local melting of the catalyst and formation of clinkers. However, when these small hot zones form next to the reactor, they can decrease the wall strength ad lead to eventual leaks of hot reactant mixtures, which lead to explosions. Special local wall control procedures have to be used to detect such local hot zones.

Another important yet unsolved problem is how to circumvent the destruction of diesel particulate filters (DPFs) due to formation of local excessive hot regions. The DPF is a multi-channeled ceramic filter. The exhaust gases flow through the porous ceramic walls and exit from the surrounding channels, while the PM is deposited in the channels and has to be periodically removed by combustion. The combustion often generates moving high temperature waves, the temperature of which may exceed in some cases the melting temperature of the DPF channels, damaging the DPF. Significant research is currently being conducted to develop regenerating procedures which avoid the destruction of the DPF by local melting.

Polymer Reaction Engineering - Historical Development, Current Vitality and Some Future Challenges

W. Harmon Ray, Chemical and Biological Engineering, University of Wisconsin - Madison, 1415 Engineering Dr., Madison, WI 53706

Abstract

This talk will describe the rise of Polymerization Reaction Engineering as a discipline leading to new catalytic and kinetic mechanisms, new chemical species, and new polymer products over the last 100+ years. Some exciting new directions that illustrate the vitality and future promise of this field will also be discussed.

94a Education Challenges and Opportunities in Pharmaceutical Engineering as An Emerging Academic Discipline

Piero M. Armenante, Otto H. York Department of Chemical, Biological, and Pharmaceutical Engineering, New Jersey Institute of Technology, 323 King Blvd., Newark, NJ 07102-1982

Pharmaceutical engineering is becoming a distinct and well recognized engineering discipline, as indicated by the rapidly growing number of programs that have recently been developed. Nevertheless, pharmaceutical engineering is still a new engineering field in academia and it is important to be able to define its core disciplines and its range of activities, especially since pharmaceutical engineering curricula and the courses that constitute them are still being identified.

In this talk, the current status of pharmaceutical engineering education is examined, with specific reference to the Pharmaceutical Engineering program which has was established at NJIT in 2002 - one of the first of such programs established worldwide. This presentation discusses what can constitute the common knowledge in this rapidly growing field and examines the education challenges and opportunities associated with the definition of the core courses, elective courses, and curriculum activities for such programs.

94b The Changing Role of Chemical Engineers in R&D in the Pharmaceutical Industry

Edward Paul, Self, 308 Brooklyn Blvd., Sea Girt, NJ 08750 and Michael P. Thien, Global Pharmaceutical Commercialization, Merck & Co., Inc., P.O. Box 2000, Rahway, NJ 07065.

The role of chemical engineers in R&D in the pharmaceutical industry has evolved over the past decades in response to changing technological and regulatory challenges. The focus of academic preparation has also evolved from an emphasis on petroleum based chemistry and continuous processes to increasing attention to the needs of the pharmaceutical industry. These needs can include preparation for the engineering of increasingly complex chemical and biological processes. One question is always present: can this product candidate be actually made on a commercial scale or will it remain a laboratory curiosity with great therapeutic potential but no practical way to bring it to market?

Examples will be provided to illustrate how chemical engineering has contributed to provide solutions to some of the challenges.

94c An Introspective of the Regulatory Influence on Drug Development

Christine Seymour, Pfizer Global Research & Development, Pfizer, MS 7040-1208, Eastern Point Rd, Groton, CT 06340

Text Not Available.

Biochemical engineering has a rich history of contributions to human health, and to the chemical engineering profession as well. From a definition and blossoming of the field in the 1960's the field radiated with the advent of rDNA in the 1980's, and apparently (?) reached a maturation phase in the 2000's. With this has come a shift away from the traditional process development and manufacturing emphasis, with altered training and funding for academics. Where are the future opportunities in biopharmaceuticals, and where should the field go? This talk will discuss current challenges in biologicals development and production, and give a perspective on the future of biochemical engineering for the biopharmaceuticals industry.

109a What Can One Do with a New Separation Process?

Edwin N. Lightfoot, Chemical and Biological Engineering, University of Wisconsin--Madison, 1415 Engineering Drive, Madison, WI 53706

Introduction

It is now clear that we have been caught napping with outmoded technology in many key areas of our economy, and the basic reason for this is our tendency toward conformity. This is a deeply imbedded human trait that may well have originated when distant ancestors left the relative safety of trees to live in savannahs. The fearsome predators there could only be kept at bay by presenting united fronts of several individuals. Loners were quickly weeded out, and most of us are still uncomfortable going our own way. The great utility of original thinkers sufficed to keep a few in the gene pool, but they have always been viewed with suspicion – and not least by CEO's and granting agencies. As a result we tend not to have effective research strategies until we are forced to develop them by major challenges. Now after a half century of a nearly free ride thanks to favorable post WWII circumstances the challenges have arrived.

As Alan Greenspan [The Age of Turbulence, Penguin 2007] has pointed out a dynamic economy is normally characterized by continuing constructive destruction: the replacement of established processes and products - or whole societies - by new and more economical ones. Some overseas competitors seem to realize this better than we.

Mark Buchanan [The Mathematical Mirror to Animal Nature, 2008, 453, 714-716] has neatly described a closely analogous situation faced by most animals: Food is not, in general, spread equally around the world. It comes in lumps. Foragers thus need a strategy for finding those lumps. So do we, and just following others is not in general an effective strategy. That is proven over and over again, by lemmings for example. And by the stubborn refusal of most industries and granting agencies to fund exploratory research.

Surely this should not be true of downstream processing, a complex mixture of separation processes that dominate production costs in one of our most dynamic industries. We find however, that new separation processes in biotechnology have been remarkably few. This is no doubt true in part because old processes, such as chromatography, are continually being improved and specialized. However, it remains a surprise to this author that true steady counter flow processes, dominant in most other areas of the chemical process industry, do not yet have a foothold. It was decided therefore to produce such a process, ideal counterflow membrane cascades, to see if a place can be found for it, and to suggest a "foraging strategy". The strategy is a simple type of emergence: a novel combinations of well established ideas.

This new process has now been reasonably well characterized and shown to give much improved separations relative to the closely related simple tangential flow membrane filters or the inefficient cascades of them reported from time to time in the reviewed literature. It remains to find significant specific applications for which this new process is competitive and to demonstrate its utility.

To do this requires a comprehensive picture of downstream processing requirements and costs for our complex industry, and this is the daunting but important purpose of this talk. Such a task is in essence a mystery, i.e., too complex for complete solution, and we must seek to define a puzzle, a problem simple enough to solve, and powerful enough to provide useful insight. Fortunately we already have a significant body of literature for guidance.

The nature of downstream processing

It is immediately clear for example that a separation trajectory can be drawn on a plot of thermodynamic activity vs. concentration for the desired species (Fig. 1), and it appears that there are three separately identifiable tasks required for success. These are usually defined as

Capture or concentration,

Fractionation, and

Purification or polishing

We shall be most interested in the second of these, but the first provides a very useful framework for our discussion. It should also be noted that it is sometimes possible to combine two or more of these tasks, but that possibility will not be discussed here.

For the conditions of most interest here the first of these three processes dominates manufacturing cost, and it in turn is dominated by simple materials handling [E. N. Lightfoot and MCMC]. It has long been known (Fig. 2, [Sherwood], and [figure author]) that cost is inversely proportional to concentration of the desired product in the feed, and it is remarkably insensitive to the specific nature of the system being investigated. Moreover, the feed is normally a very complex multicomponent mixture and not well suited to processes such as membranes which produce only binary or pseudo-binary separations.

From our present standpoint capture is primarily of interest because it suggests that the major processing cost in the remaining two tasks is yield loss, and that the primary measure of success is yield relative to increase in purity. It is also suggested in this talk that fractionation and polishing can be viewed separately.

The last major insight behind this discussion is the analogy between membrane filtration and the various processes used for isotope fractionation. It suggests that the Dirac value function [see BSL, Ex. 23.1-3, p. 735] may be a useful tool for comparing alternate fractionation processes. This approximation will be examined in the course of discussion.

Finally specific examples will be examined to firm up the discussion and to test the various approximations made in earlier phases of the work. These are still to be selected at the time of writing.

109b Advice to a Young Bioseparations Scientist/engineer: Thoughts from a Career

Georges Belfort, Chemical and Biological Engineering, Rensselaer Polytechnic Institute (RPI), 110 8th St., Troy, NY 12180

In this retrospective, we will review some key discoveries and developments that form the foundation of pressure-driven synthetic membrane processes and their applications to bioseparations. We will also place them in the context of the author's career and look for rules or lesson on how to develop ones career in science and engineering. Specifically, we will discuss the early work of describing the application of the first membrane modules for biotechnology by Madsen, Nielson and Olson at the Danish Sukkerfabriken in the 1970s. Early attempts by Thomas and Dresner at Oak Ridge to deal with fouling are most instructive and laid the foundation for most analyses by groups in the US and Europe. Others have followed Chao Chu's ideas and patents at AMF in the 1980s that led the way to develop charged repulsive membranes that are again the topic of much research. Within the presentation, a track of the author's career will be used to emphasize how young scientists could perhaps optimize their efforts in setting up a scientific career.

109c Membrane Bioseparations – Look How Far We've Come!

Andrew Zydney, Chemical Engineering, The Pennsylvania State University, 160 Fenske Laboratory, University Park, PA 16802

Membranes are used extensively throughout the pharmaceutical and biotechnology industries for a wide range of separations. This talk will briefly trace the historical development of some of the key membrane bioseparation processes and will then focus on recent developments in membrane science and technology that are likely to have a significant impact in the coming years. Emphasis will be placed on showing how our improved understanding of the underlying transport phenomena have led to exciting new applications of membrane technology for bioseparations. Two examples will be highlighted: the development of electrically-charged membranes for protein purification and the use of ultrafiltration systems for the separation of plasmid DNA isoforms based on differences in flexibility. The use of electrically-charged membranes significantly increases the retention of like-charged proteins, providing the potential for highly selective separation of proteins with very similar properties. Membranes are particularly attractive in the large-scale purification of plasmid DNA for gene therapy and vaccines, overcoming the low capacities of chromatographic techniques. These developments have opened up exciting new opportunities for membrane bioseparations in the purification of high value therapeutics.

109d History of Insulin Purification

Warren MacKellar, Technical Service - Biosynthetics, Eli Lilly and Company, Lilly Corporate Center, Indianapolis, IN 46285

Text Not Available.

Separations Challenges in Aqueous Separations

Michael Ladisch, Purdue University, Laboratory of Renewable Resources Engineering (LORRE), Purdue University, Potter Engineering Building, 500 Central Drive, West Lafayette, IN 47907-2022

Abstract

The separation of biomolecules from aqueous solutions has been a major area of emphasis in the production of biopharmaceuticals and other high value products. Numerous creative and cost-effective solutions have been developed and integrated into biopharmaceutical manufacturing facilities over the last 20 years. A new set of challenges is now emerging as technologies for cost-effectively producing small molecules from cellulosic materials are being developed. These small molecules range from ethanol and related alcohols that could be used as biofuels, to fermentation products that may also be useful as precursors for the chemical industry. All have a common denominator in that they are produced in aqueous environments, and require cost-effective methods for their recovery and purification at relatively dilute concentrations. This paper addresses some of the challenges in processing fermentation-derived products from aqueous fermentation broths, as well as the role of separations in achieving cost-effective, aqueous-based manufacture of small molecules. Examples will include recovery of ethanol from fermentation broth, biobutanol, and monosaccharides that may be derived from non-food, renewable resources that have a low carbon footprint, and would develop from platforms that provide inexpensive, fermentation derived, precursor molecules. These renewable resources include cellulosics such as wood, corn stover, and residue obtained from the processing of corn grain to ethanol. The impact of liquid/solid separations, novel adsorption methods, and chromatographic separations will be discussed.

143a The Role of Mathematical Programming in the Evolution from Planning and Scheduling towards Enterprise-Wide Optimization

Ignacio E. Grossmann, Center for Advanced Process Decision-making, Dept of Chemical Engineering, Carnegie Mellon University, 5000 Forbes Avenue, Pittsburgh, PA 15213

Dating back to the 60's the petrochemical process industry has been a pioneer in the application of mathematical programming, mostly in the form of linear programming, for production planning, despite the nonlinearities that are encountered in process models. In contrast, in process scheduling the industry has historically relied on heuristics and ad-hoc spreadsheet tools. However, in the last 10-15 years, given important advances in mixed-integer linear and nonlinear programming, there has been a strong move towards the use of mathematical programming in the area of scheduling. Furthermore, these advances have also promoted the holistic and more ambitious view of optimizing entire supply chains, consisting of suppliers, manufacturing plants and distribution centers, which has led to the goal of performing enterprise-wide optimization. In this presentation we discuss some of these developments, and emphasize future challenges that need to be overcome to realize the vision of Enterprise-wide Optimization. We specifically discuss challenges related to modeling, multi-scale optimization, optimization under uncertainty and algorithmic and computational developments that need to be overcome to make enterprise-wide optimization a reality. It is our belief that a multidisciplinary approach from chemical engineering and operations research is required to make significant progress in this area through the combination of advances in modeling of scheduling, planning and supply chains, and mathematical programming techniques for mixed-integer, disjunctive, global optimization and stochastic programming. We provide several examples of recent successful industrial applications in planning and scheduling of petroleum and chemical facilities that are clearly pointing to achieving the goal of Enterprise-wide Optimization.

Web Page: egon.cheme.cmu.edu/

143b Production Planning and Scheduling, Computer Integrated Manufacturing and Supply Chain Management

Shinji Hasebe, Department of Chemical Engineering, Kyoto University, Katsura Campus, Nishikyo-ku, Kyoto, 615-8510, Japan

There are long history of the studies on production planning and scheduling. In this talk, the trend of the research on production planning and scheduling, computer integrated manufacturing, and supply chain management are discussed from several directions. First, these problems are classified according to the constraints and objective functions of each problem. The system boundary is one of the key issues which distinguish the problem formulation. Then, these problems are classified according to the treatment of production demand. How to express the future demand and its uncertainty is used to classify the problems. Finally, the effect of advanced optimization methods on the formulation of the problem is discussed.

143c Large-Scale Nonlinear Programming: An Integrating Framework for Enterprise-Wide Dynamic Optimization

Lorenz T. Biegler, Department of Chemical Engineering, Carnegie Mellon University, 5000 Forbes Ave., Pittsburgh, PA 15213

Integration of real-time optimization and control with higher level decision making (scheduling and planning) is an essential goal for profitable operation in a highly competitive environment. While integrated large-scale optimization models have been formulated for this task, their size and complexity remains a challenge to many available optimization solvers. On the other hand, recent development of powerful, large-scale solvers leads to a reconsideration of these formulations, in particular, through development of efficient large-scale barrier methods for nonlinear programming (NLP). As a result, it is not unrealistic to solve NLPs on the order of a million variables, for instance, with the IPOPT algorithm. More recently, an NLP sensitivity extension to IPOPT quickly computes approximate solutions of perturbed NLPs. This allows on-line computations to be drastically reduced, even when large nonlinear optimization models are considered. These developments are demonstrated on dynamic real-time optimization strategies that can be used to merge and replace the tasks of (steady-state) real-time optimization and (linear) model predictive control. We consider several process case studies to illustrate these concepts.

143d The Evolution, Growth and Future Perspectives of Operations in Process Systems Engineering

G. V. Reklaitis, Chemical Engineering, Purdue University, Forney Hall of Chemical Engineering, Room 1019, 480 Stadium Mall Drive, West Lafayette, IN 47907-2100

The initial focii for process systems engineering were in process control and design. While initially confined to regulatory control of units and process trains, the former area has expanded to encompass the additional decision layers of the operational hierarchy from process monitoring through real time optimization and most recently supply chain and even enterprise-wide operational management. In this paper we will review this evolution both in terms of the technical developments and the supporting conferences, meetings and publications which served to bring together the sector of the pse community with interests and needs in process operations.

154a Computational Statistical Mechanics: Past, Present, and Future

Glenn H. Fredrickson, Chemical Engineering, University of California, Santa Barbara, Engr. II, Room 3357, UCSB, Santa Barbara, CA 93106-5080

I will discuss the evolution of numerical approaches to simulating the equilibrium and time dependent properties of simple and complex fluids. A particular emphasis will be placed on rapidly evolving field-theoretic methods, their connection to computational advances in continuum mechanics, and on emerging techniques for bridging length and time scales.

154b Perspectives on the Past, Present and Future Advances in Computational Chemistry and Their Application to Catalysis

Matthew Neurock, Chemical Engineering, University of Virginia, 102 Engineers' Way, PO Box 400741, Charlottesville, VA 22904

We review the advances that have taken place in the development of electronic structure theory as well as atomistic simulations over the past century and describe their application to modeling catalytic systems. In addition, we discus current challenges in the field as well as future developments.

154c Thin Films and Nanostructures of Hard Materials: Progress Made and Current Challenges In the Computational Modeling of Their Processing and Function

Dimitrios Maroudas, Department of Chemical Engineering, University of Massachusetts, 686 N Pleasant Street, Amherst, MA 01003

In this presentation, we discuss the progress that has been made, as well as the current challenges, in the computational modeling of thin-film and nanostructured forms of hard materials, such as metals and semiconductors, which are used broadly in engineering applications including microelectronics and nanofabrication. Multiscale and hierarchical modeling approaches are outlined based on first-principles density functional theory calculations, large-scale molecular-dynamics simulations, coarse-grained Monte Carlo methods, computational continuum mechanical techniques, and nonlinear analysis of dynamical systems. The problem areas presented include semiconductor surface chemistry and transport, stress- and plasma-induced structural transitions, mechanical behavior of small-volume structures, and morphological response to the combined action of multiple external fields.

154d Polymer-Induced Drag Reduction of Turbulent Flows: What Have We Learnt from Direct Numerical Simulations ?

Antony N. Beris[1], Gaurab Samanta[1], and Kostas D. Housiadas[2]. (1) Chemical Engineering, University of Delaware, 150 Academy street, Newark, DE 19716, (2) Mathematics, University of the Aegean, Karlovasi, Samos, 83200, Greece

The phenomenon of polymer-induced drag reduction describes the effect that even minute quantities of a high molecular weight polymer can have, of the order of ppm by weight, when added to a low molecular weight solvent, such as water or crude oil, in reducing considerably the turbulent drag. Despite a voluminous amount of research from its original discovery in the late 40s by Mysels and Tomms, unresolved issues still remain. Here I will describe the recent progress achieved thanks to Direct Numerical Simulations (DNS) of the turbulent channel flow of a dilute polymer solution modeled from first principles with various constitutive equations such as (FENE-P), (Giesekus), or others.

The main effect of viscoelasticity is shown to be the strengthening of the largest size turbulent structures which become much more coherent with a dynamics that changes at an appreciably lower rate than for the equivalent Newtonian structures. Our parametric study strongly suggests that this feature develops due to an enhanced resistance to extensional deformation induced due to viscoelasticity and it results to a lower energy transfer from the wall to the turbulent core, thus explaining the drag reduction. Recent developments in numerical methods allowed us to obtain accurate and stable simulations up to highly drag reduced (more than 60%) turbulent channel flows of dilute polymer solutions. The most recent data confirm earlier results in our group at lower drag reduction values whereby the primary mechanism for drag reduction is the decreased intensity of the wall eddies which is the result of a significant increase to extensional deformations contributed by the polymer additives, exactly as proposed earlier by Metzner and Lumley.

Recent work has been able to more systematically investigate the changes to the flow structure affected due to the polymers, and in particular to the coherent structures, using Karhunen-Loeve (KL) or Proper Orthogonal Decomposition (POD) analysis of the DNS data. This demonstrated a dramatic decrease in the K-L dimension of the flow (an order of magnitude) with viscoelasticity versus the Newtonian flow. Moreover, the time-dependent dynamic analysis of the solution in terms of the K-L modes a reduction in the velocity data has confirmed the increased time scales encountered with viscoelasticity but also the sensitivity of the polymer conformation data and stress to many scales of the velocity field. It also underscores the tremendous difficulty underling any effort towards a low-dimensional modeling of turbulent flows, given the large extent of scales of length and time characterizing turbulent dynamics.

Dynamic Models of Biological Systems. A Hotbed for Nonlinear Analysis.

By

Doraiswami Ramkrishna
School of Chemical Engineering
Purdue University
West Lafayette, IN 47907

Chemical engineers have exploited nonlinear analysis of dynamical systems in diverse ways during the last century in the fields of chemical reaction engineering, fluid mechanics and transport processes. The objective of this presentation is to call attention to numerous topics in biological applications that represent opportunity areas for nonlinear analysis. The essential requirement of course is the availability of good dynamic models of such complex systems. While conservation principles and good constitutive supplements have proved sufficient for modeling of physical systems, biological systems add substantially to the challenge through features unique to them.

The unique features of biological systems are related to their capacity, acquired through exposure to eons of evolutionary pressures, to respond to their environment in ways that promote their survival. The consequence is an enormous increase in the diversity of responses of the system that modeling must necessarily represent within a dynamic framework. This requirement is enforced by the uncertainty surrounding our assessment of the organism's evolutionary experience that cannot be resolved without the institution of an interactive program of experiment and theory growing with contributions from each other. From this perspective, the opportunity in this area is not to be viewed simply as a Mecca for modelers for solely intellectual contentment. Rather, it will promote understanding of complex systems through the development of new experimental protocols that will provide a deeper setting for uncovering intriguing behavior. Some examples will be discussed.

The modeling framework involves viewing metabolism as a combination of so-called elementary modes[1] along which external substrates are taken up for metabolic activity relating to the production of various intracellular metabolites, biomass and fermentation products. The elementary modes represent multiple options for the organism. While steady state methodologies postulate a fixed distribution of external substrate among various elementary modes, the cybernetic approach of Ramkrishna and coworkers postulates a distribution *controlled* so that global objectives such as maximization of growth rate or uptake rate of substrate is accomplished. In particular, the hybrid model (Kim et al., 2008[2]; Song and Ramkrishna[3]) postulates cybernetic control of substrate uptakes through elementary modes so that a global objective such as the maximization of

---

[1] S. Schuster, D. Fell, T. Dandekar, *Nature Biotech.*,**18**, 326-332, 2000

[2] J. Kim, J. Varner, D. Ramkrishna, *Biotechnol. Prog.*, in press.

[3] Song, Hyun-Seob and Doraiswami Ramkrishna, "Reduction of a Set of Elementary Modes by Yield Vector Analysis," *Biotechnol & Bioeng.*, in press.

substrate is realized, while internal metabolism in each mode occurs at steady state with respect to the uptake rate through that mode. The resulting dynamic models are equipped to handle large metabolic networks including the role of regulatory processes.

As an example, anaerobic growth of *E coli* is considered in a mixture of glucose and pyruvate. The organism can consume the two substrates either sequentially or simultaneously depending on the concentrations of the two substrates and the organism's preculturing. The metabolic network used for the modeling is shown in Figure 1.

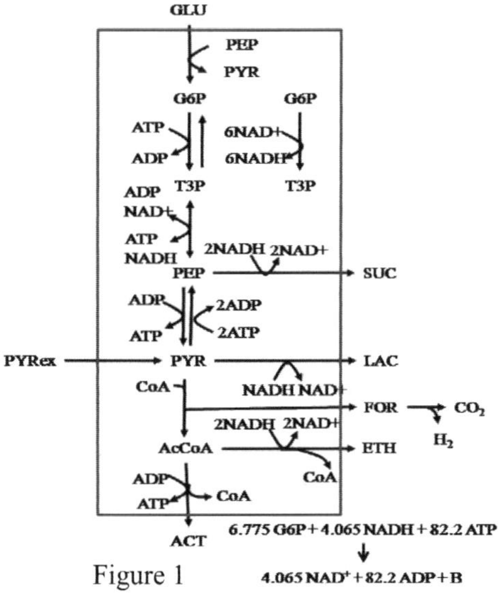

Figure 1

The batch experiments of anaerobic *E.coli* (strain GJT001), obtained from Ka-yiu San's group of Rice University, were implemented to identify the kinetic parameters in the hybrid model.

The reduced metabolic network of anaerobic *E.coli* is shown in "Fig. 1". All biochemical reactions in the reduced network is originated from Young's reduced network.[4] However, some modifications were made based on the information in the website of B.Ø.Palsson's group[5], since the pyruvate uptake and other reactions must be considered additionally.

Applying pseudo steady state assumption to internal species in "Fig. 1"., total 19 elementary flux modes were identified using standard software.[6]

Among a total of 19 elementary modes, 15 were eliminated to finally yield the four elementary flux modes displayed in "Table 1" - accomplished by fitting the cybernetic model parameters to the experimental data on all fermentation products, substrates, and biomass concentrations in batch cultures. To simplify the parameter fitting process, five modes were pre-eliminated as they did not appear to be competitive.

Table 1. Major elementary flux modes of anaerobic *E. coli*

I	Elementary flux mode ($\mathbf{z}_i$)
1	35.76 GLU = 57.96 FOR + 31.01 ACT + 26.95 ETH
2	150.76 PYR = 133.42 FOR + 133.42 ACT
3	13.55 GLU + 66.62 PYR = 80.17 FOR + 75.43 ACT + 4.74 ETH
4	8.81 GLU + 85.59 PYR = 89.65 FOR + 89.65 ACT

The four modes shown in Table 1 were treated as four competing major reactions of metabolism. Each reaction rate and enzyme synthesis were assumed to satisfy Michaelis-Menten kinetics, and those reactions were regulated by cybernetic variables to maximize the total carbon uptake rate. The model was simulated for elucidating the steady state picture as a function of dilution rate and glucose fraction (γ) in the feed. A narrow window of dilution rates was found displaying steady state multiplicity with hysteresis.

[4]J. Young, K. Henne, J. Morgan, A. Konopka, D. Ramkrishna, *Biotech. & Bioeng.,* **100**, 542-559, 2008.
[5]http://gcrg.ucsd.edu/downloads/index.html
[6]http://pinguin.biologie.unijena.de/bioinformatik/networks/metatool/new_metatool/new_metatool.html

This result was verified experimentally as shown in Figure 2. A total of 6 different shift-ups and shift-downs of dilution rates were made to check the predicted hysteresis with 3 steady states.

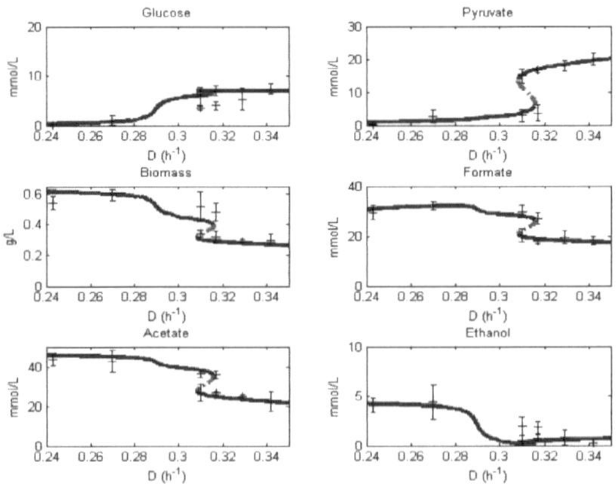

Fig. 2. Steady-state plots of continuous *E. coli* cultures at different dilution rates. (Solid line : Stable steady state plot, Dashed line : Unstable steady state plot, Error bars : 90% confidence interval of experimental data)

For higher concentration fractions γ of pyruvate, the steady state multipli-city was even higher as shown. Thus in Figure 3 are displayed as many as 5 steady states. Experimental verification is currently in progress.

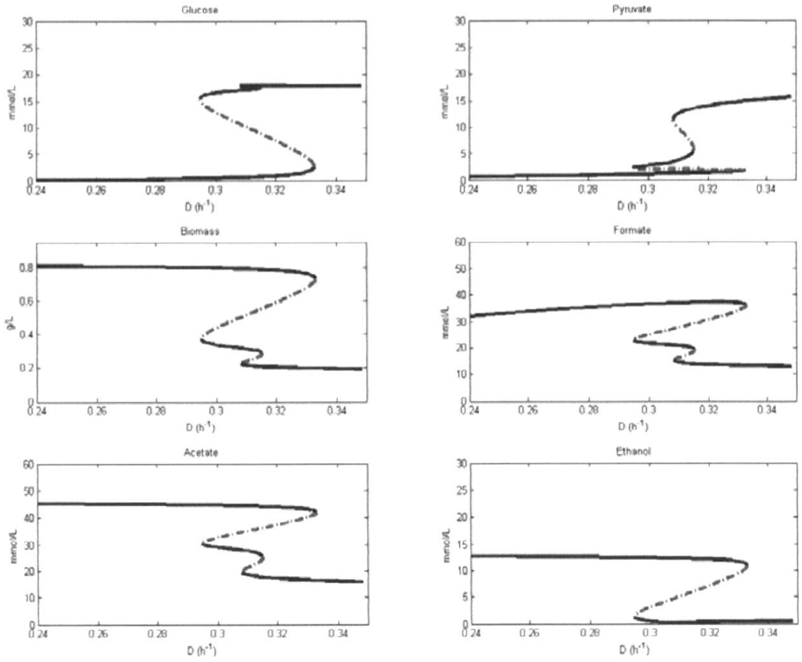

Figure 3. Steady state multiplicity in continuous reactors fed with a mixture of glucose and pyruvate.

The model also predicts oscillatory behavior (see Figure 4below) at low dilution rates with glucose-only feed. (Such dilution rates are indeed encountered in water treatment). With mixed feeds such Hopf bifurcation behavior becomes even more possible. These results require verification.

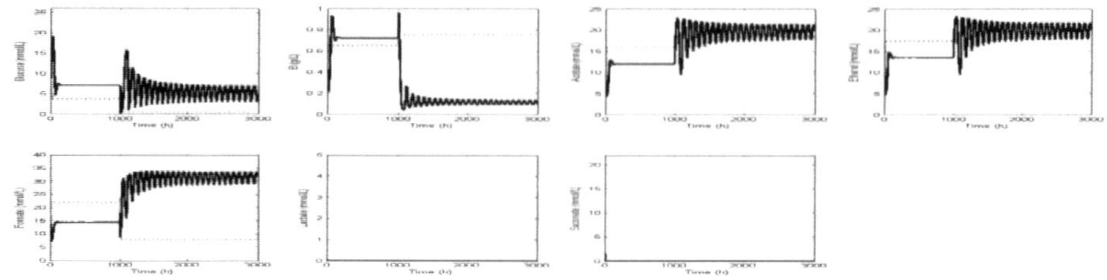

Figure 4. Oscillatory Behavior predicted by the dynamic cybernetic model

Experimental support for the above multiplicity and oscillatory features should not be viewed as a mere test of the implications of some specific nonlinearity. It is a consequence of manipulation of elementary modes by the organism in response to environmental changes. Biological systems have an inherence propensity for such nonlinear behavior because of metabolic regulation.

154f Paint-by-Numbers: Pattern Formation In Two-Dimensional Sheets of Cells

Stanislav Y. Shvartsman, Department of Chemical Engineering and Lewis-Sigler Institute for Integrative Genomics, Princeton University, Engineering Quadrangle, Princeton, NJ 08544

One of the basic mechanisms responsible for the formation of three-dimensional organs during embryogenesis relies on the regulated folding of two-dimensional sheets of cells. This process is driven by the spatially nonuniform and dynamic distribution of multiple chemical components (products of gene expression). Some of the key questions in this class of problems are related to the number of involved genes and to the diversity and dynamics of their expression patterns. I will present the results of our experimental and computational work that explores these questions in fruit fly development. I will discuss the mathematical and computational issues associated with the combinatorial construction of complex patterns from a small number of building blocks and the dynamics of piecewise linear models of patterning. Finally, I will compare and contrast models of biological pattern formation with models of purely chemical systems.

Three Phases of Chemical Reaction Engineering
AIChE Philadelphia, November 17, 2008
James Wei, Princeton University

Abstract

Chemical Reaction Engineering has evolved in the last century, and can be divided into three phases. The first phase of Empirical Craft is exemplified by textbooks of descriptive industrial chemistry for different industries, to codify decades of empirical experiences. When the influence of unit operations began to be felt, a series of textbooks in chemical reaction engineering emerged in the second Science-Based phase, and written to be much more useful for process improvements and innovations, as well as for operations. Complex kinetics and reactor problems had to be simplified so that they can be solved by analytical and graphical methods. When computers became more powerful and available, much more complex and realistic aspects can be considered, which led to the third Computer-Based phase of textbooks. The dominance of Commodity chemicals has to make way for newer Specialty industries that have smaller volumes and shorter product life cycles, which pose new challenges to reaction engineers.

Chemical Reaction Engineering has gone through many phases, which is reflected in the dominant textbooks of that phase. The construction and operation of chemical reactors are as old as the first hearth for heat and light, and the first fermenter to make beer and wine, which developed over the ages and improved over generations. The development of modern chemical reactors were accelerated by the large scale production of chemicals such as the LeBlanc process for making soda ash in 1791, the introduction of nitroglycerin by Sobrero in 1846, and the invention of synthetic dye by Perkin in 1856. Books are needed to instruct a newcomer how to design and build appropriate chemical reactors, and how to operate them successfully. For an author to master the most interesting and proprietor aspects of an industry and the reactors, it would be necessary to have the insider knowledge of an employee, or a regular consultant and collaborator. It is more manageable when a chief editor brings a team of experts from different industries to produce a book.

Phase I. Empirical Craft: Descriptive Industrial Chemistry

In this early phase that began before the founding of the AIChE in 1908, the textbooks described the successful industrial technologies which were evolved over several generations of development and empiricism. Some of the leading textbooks of this genre and their descendants were:
1898 F. H. Thorpe, "Outline of Industrial Chemistry"
1928 E. R. Riegel, "Industrial Chemistry"

1935 P. H. Groggins, "Unit Processes in Organic Synthesis"
1945 R. N. Shreve, "Chemical Process Industries"
These books mainly tell you how things are done, without elaborations on scientific explanation on why they are done this way, and why other ways would not work. The starting point may be a verbal or a drawn flowchart of the various streams in the manufacturing of a product such as sulfuric acid, which contains reactors among various auxiliary equipments. There is description of the raw material sources and feedstock preparation, the operating conditions such as temperature, pressure, and residence time, the product distribution from the reactor, and finally the market for the product.

If you are entering employment in a sulfuric acid plant, this type of book can serve as orientation of the establishment that you have joined. However this book is short on the factors that compel you to design and operate the reactor this way, and how you may make contributions by taking incremental improvements in efficiency and yield, to say nothing of radical breakthroughs in innovations. If you are an engineer at a developing nation who has just received an order to build a sulfuric acid plant, this book can suggest ideas on what to do as a start, but without the details necessary to design, build and operate an efficient plant. What can this book do for students of chemical engineers who are planning to work for another branch of the chemical industry, such as petroleum refining? Can the lessons learned here be applied to manufacture other products under current production, or brand new discoveries and inventions. This mode of teaching is appropriate in a static environment when you are asked to do the same thing as your predecessors in the same way, but it suffers in a dynamic environment as it cannot be easily generalized to a different or a new industry. It does not provide you with knowledge and skills that are generally useful in many other applications, and to deal with dynamic new developments.

Progress was very slow in this period. It took many pioneers over the course of many decades to consolidate these isolated cottage technologies into the learned discipline of reaction engineering, which is a paradigm of recognized and systematic body of knowledge in the university curriculum, accredited by professional societies, recognized topics of research, presented in learned society meetings, and published in archival journals.

<u>Phase II. Science-based Reaction Engineering: petroleum and petrochemicals</u>

The coming of Unit Operations made a profound change in the way our profession looked at research and teaching as sources of innovation and improvement in chemical engineering. This paradigm shift looked for common threads among various branches of the chemical industries for knowledge that are generally useful, and led to Unit Operations in the book "Principles of Chemical Engineering" by Walker-Lewis-McAdams in 1923. The subjects of inquiry were no longer the individual industries, but were the principal physical operations that take place in all chemical plants, such as pumping fluids, heating and cooling, distillation and absorption. Once the design and operation of individual equipments are understood, any chemical plant can be designed as an orderly synthesis of well understood constellation of parts. However this revolution has left out a most important component - the chemical reactor which is the heart of making useful things.

We see in the next generation textbooks in reaction engineering, which adopted the same unifying approach, and were organized around different types of kinetics and reactors, instead of around different industries:
1947 O. Hougen and K. Watson, "Chemical Process Principles, part 3, Kinetics and Catalysis"
1956 J. M. Smith, "Chemical Engineering Kinetics"
1962 O. Levenspiel, "Chemical Reaction Engineering"
1963 H. Kramers and K. R. Westerterp, "Elements of Chemical Reactor Design and Operation"
1966 K. Denbigh, "Chemical Reactor Theory"
The topics in Hougen and Watson started with individual and networks of chemical reaction kinetics, with considerations of reaction order and Arrhenius relations to temperature, and of thermodynamics and equilibrium. Two chapters have the titles of "Mass and Heat Transfer in Catalytic Beds" and "Catalytic Reactor Design". We find here concepts that have become staples: the Thiele modulus of diffusion in catalyst pellets, longitudinal diffusion in flow reactors, graphical calculation of reactor sizes, pressure drops in granular beds, and optimum reaction temperatures.

The most dynamic industries at that period in rapid development and growth were petroleum refining and petrochemicals, where catalytic reactions held the center place, so that most of the classic examples and homework problems were taken from that industry. This is not surprising since many leading professors were also consultants to these companies, drawn by the challenge of unsolved and important problems of feasibility, efficiency, scale-up, and safety. The commodity chemical industries have good problems for chemical engineers to solve, as the scales are enormous so that even one percent improvement in efficiency and yield would mean millions of dollars, and the product cycles are long enough so that process improvements are worth doing. In this phase, chemical reaction engineers concentrated in analyzing critical steps, in quantitative problems of optimal design and operation, and of stability, which can be solved by analytical methods, approximation methods, and simple graphic methods. Chemical engineers have also developed a number of textbooks in applied mathematics to solve their more complex problems:
1923 Hitchcock and Robinson "Differential Equations in Applied Chemistry"
1939 T. K. Sherwood and C. E. Reed "Applied Mathematics in Chemical Engineering"
1947 W. R. Marshall and R. L. Pigford "The Application of Differential Equations to Chemical Engineering Problems"
1962 L. Lapidus "Digital Computation for Chemical Engineers"
1966 N. R. Amundson "Mathematical Methods in Chemical Engineering: Matrices and Their Applications"
The first nine chapters of Sherwood and Reed are about treatment of data, mathematical formulation of physical problems, and analytical solutions to ordinary and partial differential equations. When the problem is too non-linear or too complex, so that

analytical solutions are not available, chapter 10 is devoted to numerical solution of partial differential equations. This is a very slow and tedious process by slide rules or hand calculators, and will be taken up by reaction engineers only if the problem is very important and cannot be done any other way.

This method of teaching gives the students the knowledge and skill to enter any industry where a chemical reactor is employed, to understand how it works and to consider improvements in operations, and to consider design improvements for the more ambitious. When a new product is introduced without any previous history of manufacture, the chemical engineer has the knowledge and tools to consider what successful types of reactors are available, what are their characteristics as well as pros and cons, and how to go about doing small scale pilot plant experiments to test feasibility and stability.

III. Computer-Based Reaction Engineering: Bio, Environmental, etc.

Many of the most skilled masters in reaction engineering relish their ability to provide analytical solutions to important industrial problems, and would look down with scorn on problems that are solved by numerical methods, such as Runge-Kutta for differential equations with highly non-linear terms. There is the intellectual pride of greater knowledge and skill needed to do analysis, in comparison with the more routine application and drudgery of numerical methods. There is also the consideration that with an analytical solution, an impartial referee can easily confirm the accuracy of the results. When the solution is in the form of an equation, an engineer can answer what-if questions such as predicting the effects of changing parameter values, even extrapolations to infinity. However with a numerical solution, a referee can check the accuracy of the results only by repeating the computation, there is little intellectual beauty to boast, and each what-if question requires a new calculation.

Should the chemical reaction engineers stay exclusively with elegant and simple kinetics and reactor problems, or should they boldly tackle the most messy and complex problems important to the industry and to the nation? It is very well that complex first-order kinetic systems under isothermal conditions can be solved analytically by the matrix method, so the heart of the mystery is laid bare. The catalytic reforming reaction in petroleum refining is approximate a system of first order reaction, which works very well under isothermal conditions. The concept of lumping also reduces complexity to the point of being manageable, but at the cost of some lost of detail - similar to data compression in images in digital cameras and sounds in iPods. But an adiabatic reforming tower has a different temperature at each bed elevation, so that a different kinetic matrix would have to be used together with changing eigenvectors and eigenvalues. But the even more important catalytic cracking is carried out as many non-linear reactions, so one can either pretend that it is (any short line is a straight line) and treat it analytically, or admit that it is not and use numerical computational methods. As long as computers were slow and expensive, an investigator has the choice of doing an approximate solution or grudgingly taking on the slow computation task of a more full fledged model. When the path is long and arduous, there is always the possibility that a computation error crept in somewhere, and the results are not accurate.

There is no mistake that using computational methods is regarded as a cop-out for a proud analyst, but I suspect that if you can do computation with Roman numerals, you will look down on people who do computation with Arabic numerals. In 1962 when I was a visiting professor at Princeton, Leon Lapidus was pioneering in using the computer, but I stuck to the high road of analysis for many more years. Half a century later, it is clear that he represented the future, and I joined his team. So a new slate of textbooks of chemical reaction engineering arrived, assuming that the students already know how to do numerical methods on a computer, or can learn it soon, and proceed to deal with much more complex and realistic problems:
1986 H. S. Fogler "Elements of Chemical Reaction Engineering"
The latest edition of this book comes with a CD-ROM in the back, which contains computer software which can be used by the students to compute a number of reactor problems which are too complex to be solved analytically.
1998 L. D. Schmidt "The Engineering of Chemical Reactors"
This book takes an additional tack by considering problems from the bio-reactors, environmental modeling, combustion, and polymerization. While there is a wealth of problems from petroleum refining and petrochemicals to explain, a similar set of meaty problems in these newer industries need to be compiled by a team of writers, each has to acquire deep expertise in an industry, after years of research, practice or consulting. This is too much to expect from a single author. Perhaps a group of experts in different industries should get together to write such a textbook.

Together with the arrival of Computation Chemistry, we have the companion Computation Chemical Engineering. The current crop of reaction engineers have no excuse to deal only with simply reactor problems that can be solved by analytical methods, as they can pile on complex systems and non-linear phenomena and still expect answers. But the price paid is also significant when you want to check the accuracy of their results, to examine the effects of changes in parameter values, and to generalize to other situations.

Future Challenges

We the chemical reaction engineers should be prepared to solve the problems of the future, instead of refighting the problems of the past. What problems would we encounter in the future? We think about the future either as gradual extrapolations of current trends in society needs and technology capabilities, as well as revolutionary leaps in either. In the 1950, the emphasis was in industrial inorganic chemistry, such as making sulfuric acid and salt. The emphasis has changed to oil refining and petrochemicals, and now to biomedical engineering, electronic processing, novel materials, and environment. In the last few turbulent years, of energy and food shortage, and of global environmental concerns, the future growth appears to be in two scales:
Commodities: petroleum, new energy sources, sustainability

Specialties: bio-processing, biomedical, electronics, advanced material
The scales in tons per year in many of the Specialty industries are much smaller by orders of magnitude than in the Commodity industries, and the products are in much earlier phases of the product cycle. The Specialty industries are more in need of product engineers who can make high speed development to put novel products to the market place, than in need for incremental improvement of mature products. There would be fewer classical products and processes that last many years, and more quick processions of products and processes. This requires an adaption of chemical reaction engineering abilities and skills in ways that remain to be worked out.

This generation of reaction engineers will have more knowledge and tool available in their practice than previous generations. They will have many new challenges, as the needs of society keeps on evolving, and I wish them success and glory.

Bioreaction Engineering: Evolution and Future Needs

Michael L. Shuler, Chemical Engineering and Biomedical Engineering, Cornell University, 270 Olin Hall, Ithaca, NY 14853

While bioreaction engineering extends into prehistory (e.g. alcoholic beverages) this talk will begin with the era when people sought to educate themselves as biochemical engineers by seeking cross-disciplinary training in chemical engineering and biology. The emphasis of this talk will be on bioprocesses with occasional mention of biophysical or biomedical applications. This initial era corresponds to the development of bioprocesses for production of antibiotics and extended into small molecule production. Simple kinetic expressions were coupled to transport issues. In the 1960's the development of immobilized enzymes inspired a new wave of analysis. Also, there began to emerge concepts of modeling cells as chemical reactors with internal structure. The 1980's saw chemical engineers embracing problems in production of biologics using recombinant DNA technology. The 1970's and 1980's saw the emergence of truly modern biochemical engineering with a much higher level of integration of molecular biology with chemical engineering principles. The advancement of mechanistic biological insight allowed the application of sophisticated engineering analysis to cellular metabolism. The field of metabolic engineering is largely dominated by biochemical engineers. Additionally the emphasis on production of therapeutic proteins resulted in development of bioprocesses based on animal cells instead of bacteria and yeast. In the last decade the development of techniques to rapidly determine the genomic content of a whole organism is changing the nature of biological discovery. The primary issue in life sciences is how to relate genomic structure to the physiology of an organism and its response to environmental changes. Another technological driver is the integration of microfluidics with cell culture. Microbioreactors have the potential to improve process optimization significantly. A key factor is whether large-scale reactor performance can be predicted with such microbioreactors. Technologies based on both genomics and microfabrication will remain significant opportunities for bioreaction engineering in the next two decades. The demarcation between bioprocess technology and biomedical applications, especially in terms of diagnostics and regenerative medicine, will become very fuzzy as cross-over applications of the same technology platforms and ideas will become increasingly common.

Catalysis and Reaction Engineering: Sites, Pellets, and Reactors

Enrique Iglesia, Department of Chemical Engineering, University of California, Berkeley, Berkeley, CA 94720

Abstract

The performance of catalysts during their industrial practice depends on processes occurring at time and length scales involved in making and breaking chemical bonds at active sites, in diffusing molecules and energy within porous pellets, and in interphase and intraphase contacting within reactor vessels. The understanding and control of these processes and of their inherent coupling have benefited from advances in experimental and theoretical methods designed to observe and interpret these phenomena with increasing accuracy, fidelity, and specificity. This presentation attempts to illustrate some of these advances using examples relevant to the refining, chemicals, and energy sectors within which many of the advances emerged and found their seminal application.

At the level of active sites within complex solids, we aim to describe how chemical reactions occur in response to local concentrations and temperatures, controlled in turn by the contacting imposed by pellets and reactors, and to structural and compositional features of catalytically-relevant inorganic structures. Structure-function relations with significant predictive rigor and the evolution of strategies for selecting and discriminating relevant elementary steps and species in complex reaction networks represent two of the essential enabling components in the remarkable progress evident within the last few decades. In particular, the structure and reactive properties of active materials are becoming more evident at the level of atomic connectivities and electronic configurations, as a result of spectroscopic and kinetic methods with greater specificity and time resolution and of theoretical tools essential for rigorous interpretations of the evidence from experimental methods. For complex mixtures and reactions, treatments have evolved from those requiring extreme lumping of species and intuitive selection of their most relevant pathways to modern methods capable of enumerating elementary reactions, assessing their contributions to overall rates and selectivities, and discriminating among mechanistic choices. Molecular speciation and theoretical methods with unprecedented accuracy and sensitivity and assessments of relevance using algorithms of increasing rigor and sophistication have been essential to recent progress.

Improvements in our descriptions of the structure of porous solids and of the dynamics of diffusion within them reflect concurrent advances in the fidelity of geometric descriptions, the accuracy of relevant molecule-surface interactions, and in the assessment of the chemical consequences of confinement for catalytic events at active sites from first-principles. Imaging of complex mesoporous structures and realistic models of their geometry, simulations of transport for interacting molecules within constrained micropores, transient methods for assessing the dynamics of transport in realistic or model solids, and theoretical treatments of the intrinsic coupling between constrained motion and chemical reactivity in the context of shape selectivity will be used to illustrate the pace and consequences of progress in this area. Finally, descriptions of contacting in vessels with multiple phases and complex hydrodynamics will be briefly examined in the context of visualization and simulation methods that include increasingly specific details about contacting and mixing and of codes that incorporate these details while benefiting from algorithmic improvements and higher computational speed.

The Evolution of Electrochemical Reaction Engineering and Future Directions

Richard Alkire, Chemical and Biomolecular Engineering, Universisty of Illinois, 600 S. Mathews Ave., Urbana, IL 61801

Abstract

Electrochemical phenomena control the existence and movement of charged species in the bulk, as well as across interfaces between, ionic, electronic, semiconductor, photonic, and dielectric materials. The existing technology base of the electrochemical field is massive and of long-standing, dating to the earliest days of chemical process industry. The pervasive occurrence of electrochemical phenomena may also be seen today in nanoscale and biological systems, in microelectronic devices, and in green processes as well as in natural systems. Consider several examples:

• Materials include metals, alloys, ceramics, ionic solids, semiconductors, membranes, coatings, colloids, conducting polymers, and biological materials including proteins and enzymes.

• Phenomena that arise include conduction, potential field effects, electron or ion disorder, electroluminescence, ion exchange, passivity, membrane transport, double layers at boundaries between phases involving free charges, osmotic flow, and electrokinetic phenomena.

• Processes that depend critically on these phenomena include energy storage and conversion, corrosion, membrane separations, electrodeposition, etching, desalination, electrosynthesis of chemicals, refining of metals, and many others.

• Products that result include microelectronic devices, sensors, batteries, fuel cells, coatings, films, metals, gases, chemicals, and ceramics.

During the early yars of the past century, each electrochemical technology was tuned to the economic and technical realities at hand, similarities among many different processes were eventually be recognized, the most significant being that, for economic reasons, large-scale electrolytic processes are invariably driven to a transport-limited rate. Therefore, the electrochemical engineering research literature of the past half century focused strongly on understanding how ohmic and mass transport processes, including the effect of hydrodynamic flow, influence the potential field between electrodes as well as the current distribution, or rate of electrochemical reaction along a surface.

All structural metals are thermodynamically unstable and corrode by virtue of local anodic and cathodic regions that are driven by the energy contained in their local environment. The annual cost of corrosion in the US has been recently estimated to be $276B, or 3.1% of the US Gross Domestic Product. The design principles for corrosion, however, are very different from those used in materials processing since, when you succeed in corrosion, nothing happens. The key requirement is to understand how failure occurs, and then to design so as to intervene.

Beginning in the 1950's, mathematical modeling of the current and potential distribution in electrochemical systems including corrosion advanced steadily at the continuum level where sophisticated simulations are by now widely used to predict behavior needed for engineering design, scale-up, optimization, and process control. Continuum codes dominate the extensive modeling literature in electrochemical systems. A wide variety of phenomena can be included with the result that models are widely used for sorting out competing effects, resolving experimental data, articulating scientific hypotheses of mechanism, measuring system parameters, and predicting behavior. Such models provide a rational basis for engineering design, optimization, and control. Generally, however, they have until only recently been based on empirical characterization of the interfacial processes that appear as boundary conditions in the transport analyses.

During the past several decades, the field has evolved rapidly based primarily on a suite of remarkable new tools that provide the ability to create precisely characterized systems for fundamental study; to monitor behavior at unprecedented levels of sensitivity, atomic resolution, and chemical specificity; and to predict behavior with new theories and improved computational abilities. These capabilities have revolutionized fundamental scientific understanding of interfacial and catalytic processes, as well as contributed to the present rapid pace of discovery of novel materials and devices where product quality is determined at the molecular scale.

To drive new electrochemical discoveries toward technology innovation, new engineering methods are needed in order to ensure product quality at the molecular scale. Recent advances in computer speed and memory, numerical algorithms, and sensor technologies indicate clearly that a systematic approach is possible that integrates scientific knowledge, intuition, experimental data, and simulations. However, today's engineering design tools are based on continuum phenomena and therefore have a blind spot at the molecular scale. In addition, the primary manipulation of operating conditions during manufacturing today occurs at macroscopic length scales. A new generation of science and engineering design methods must emerge to integrate discoveries, concepts, theory, and experimental data with process engineering in order to design and control future multiscale electrochemical systems.

The challenges to building such tools include uncertainties in the physicochemical mechanisms as well as the values of thermodynamic and kinetic parameters, complexities in the simulation of model equations that can span a wide range of time and length scales, lack of manipulated variables and direct measurements of most properties at the nanoscale during processing, and the inapplicability of most existing systems tools to address systems described by noncontinuum and dynamically coupled continuum-noncontinuum models. These challenges specify the requirements for next-generation tools needed for a systematic approach to the design and control of electrochemical systems from molecules to devices.

The ability to use multiscale, multi-phenomena numerical simulations to achieve precise, quantitative understanding at new levels of magnitude, sophistication, and completeness offers a challenge for electrochemical engineering which, when met, will bring enormous benefits through rapid innovation as well as improvements in existing technological applications.

174a Past, Present and Future of Metabolic Engineering

Eleftherios Terry Papoutsakis, Dept. of Chemical Engineering & Delaware Biotechnology Institute, University of Delaware, 15 Innovation Way, Newark, DE 19711

As a distinct discipline, Metabolic Engineering (ME) was not conceived until the early 1980s, and became more extensively understood starting in about 1992. After the formative years of extensive computational and experimental network and pathway analysis, the era of genome sequences and more broadly experimental and computational genomics had arguably the biggest impact in accelerating progress and creating new platforms and paradigms for ME. We are now at crossroads as to of the evolution and impact of the field: pathway analysis of all kinds (in vivo, in vitro and in silico), synthetic biology of the deterministic, small-scale kind, chemical-biology or library-driven forced diversity, shuffling of genes and genomes, and in vivo forced evolution, are now at the center of the much broader field, which started as a rather lonely activity among some pioneers in chemical engineering, but has been now embraced by the whole community in life sciences, and beyond.

174b Past, Present and Future of Biocatalysis

Douglas S. Clark, Chemical Engineering, UC - Berkeley, Dept of Chemical Engineering, Berkeley, CA 94720 - 1462

Text Not Available.

174c Past, Present and Future of Drug Delivery

Nicholas A. Peppas, Biomedical and Chemical Engineering, University of Texas at Austin, 1044 Camino La Costa, apartment 1070, Austin, TX 78752

Text Not Available.

174d Past, Present and Future of Cellular Engineering

Larry V. McIntire, Georgia Tech, Dept of Biomedical Engineering, Atlanta, GA 30332

Text Not Available.

192a A Century of Chemical and Physical Adsorption to Characterize Catalysts

Burtron H. Davis, Center for Applied Energy Research, University of Kentucky, 2540 Research Park Drive, Lexington, KY 40511

The adsorption isotherm and accurate measurements of of the amount of adsorption had occurred by 1910. It remained for Langmuir to introduce the concept of the "checkerboard" surface and the adsorption of an atom on the board, presumably by a chemical reaction. Taylor recognized that the surface may not be uniform and introduced the term "active site" for the special ones that could cause catalysis and contrasted slow and fast adsorption. In the 1930s, the work of Emmett and Brunauer, and then Brunauer, Emmett and Teller, utilized the adsorption of nitrogen at its boiling temperature to obtain an accurate measure of the total surface area. Emmett and Brunauer also used a combination of chemical and physical adsorption to identify the total surface area as well as the fraction of surface occupied by promoter materials. Eischens in the 1950s introduced the infrared technique to measure the nature of the adsorption on metallic and acidic catalytic sites and initiated the use of instrumental techniques for these measurements. The techniques used to effect the measurement of these values have advanced to where today automated instrumentation does the measurements but still utilizes these basic techniques to identify and quantify catalytic sites.

192b Surface Science and Catalysis of Mixed Metal Oxides

Israel E. Wachs[1], Kamalakanta Routray[2], and Taejin Kim[2]. (1) Lehigh University, Department of Chemical Engineering, Iacocca Hall, 111 Research drive, Bethlehem, PA 18015-4791, (2) Chemical Engineering, Lehigh University, 7 Asa Drive, 117 Sinclair Lab, Bethlehem, PA 18015

Introduction

Bulk and supported vanadium oxide catalysts find wide application as oxidation catalysts, but many fundamental questions are still unresolved about these important catalyst systems. To better understand the similarities and differences between bulk and supported vanadium oxide catalysts, their surface catalytic active sites were chemically probed with methanol oxidation.

Experimental

The supported vanadium oxide catalysts were synthesized by incipient wetness impregnation and the bulk metal vanadate catalysts were synthesized by co-precipitation. Raman spectroscopy confirmed that the supported vanadium oxide catalysts only contained a monolayer of surface VOx species and that the bulk metal vanadates were phase pure without the presence of excess crystalline V_2O_5. The bulk and supported vanadium oxide catalysts were chemically probed with CH_3OH-Temperature Programmed Surface Reaction (TPSR) spectroscopy and steady state CH_3OH oxidation. The TPSR experiments provide the surface kinetics for the methanol oxidation rate-determining-step (k_{rds}) that involves cleavage of the C-H bond of the surface methoxy (CH_3O^*) intermediate and the number of catalytic active sites (Ns). The corresponding steady state methanol oxidation studies, in conjunction with the CH_3OH-TPSR data, allow for determination of the methanol oxidation turnover frequency (TOF) and CH_3OH equilibrium adsorption constant (K_{ads}). For the supported vanadium oxide catalysts, the number of catalytic active sites was taken as the exposed surface VOx species. For the bulk metal vanadate catalysts, the number of catalytic active sites was determined by CH_3OH chemisorption .

Results and Discussion

Supported Vanadium Oxide Catalysts. The surface reactivity of the supported vanadium oxide catalysts for breaking of the C-H bond of the surface CH_3O^* intermediate was determined by CH_3OH-TPSR and found to be very similar for all the supported vanadium oxide catalysts (similar k_{rds}). Furthermore, crystalline V_2O_5 also exhibits comparable reactivity for decomposition of surface CH_3O^* to HCHO and H^*. This suggests that the kinetics for the decomposition of surface methoxy to formaldehyde is characteristic of vanadium oxide sites. The redox TOF values of the supported vanadium oxide catalysts, however, vary by a factor of $\sim 10^3$ as the underlying support is varied. This difference is not related to structural changes of the surface VOx species since all the catalysts contain monolayer coverage and are dominated by polymeric surface vanadium oxide species. The origin of the dramatic variation in reactivity with the specific oxide support are, thus, related to the dissociative chemisorption of CH_3OH that involves breaking of the O-H bond and is reflected by the equilibrium adsorption constant (K_{ads}). The TOF dependence on K_{ads} is found to vary linearly for the different supported vanadium oxide catalysts as shown in Figure 1.

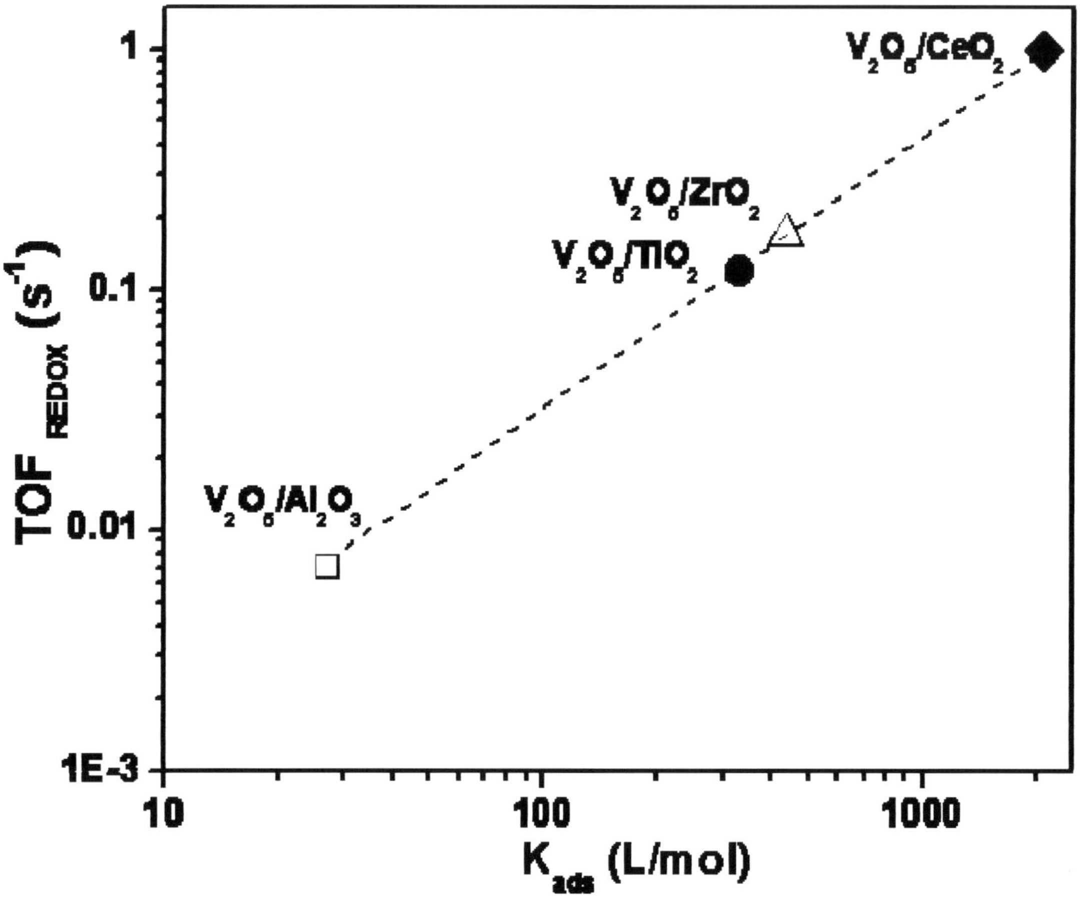

Figure 1. TOF vs. K_{ads} for methanol oxidation over supported vanadium oxide catalysts.

Bulk Metal Vanadate Catalysts. It has been proposed in the literature that the bulk V=O bond controls the reactivity of bulk metal oxide catalysts. To investigate this hypothesis, the methanol oxidation TOF values were compared with the bulk V=O bond lengths of a series of bulk metal vanadate catalysts as shown in Figure 2. The bulk V=O bond length is reflected by the Raman band position with shorted bond lengths corresponding to higher cm^{-1} values. The same relationship was also found when TOF was plotted against the V=O bond lengths determined by X-ray crystallography.

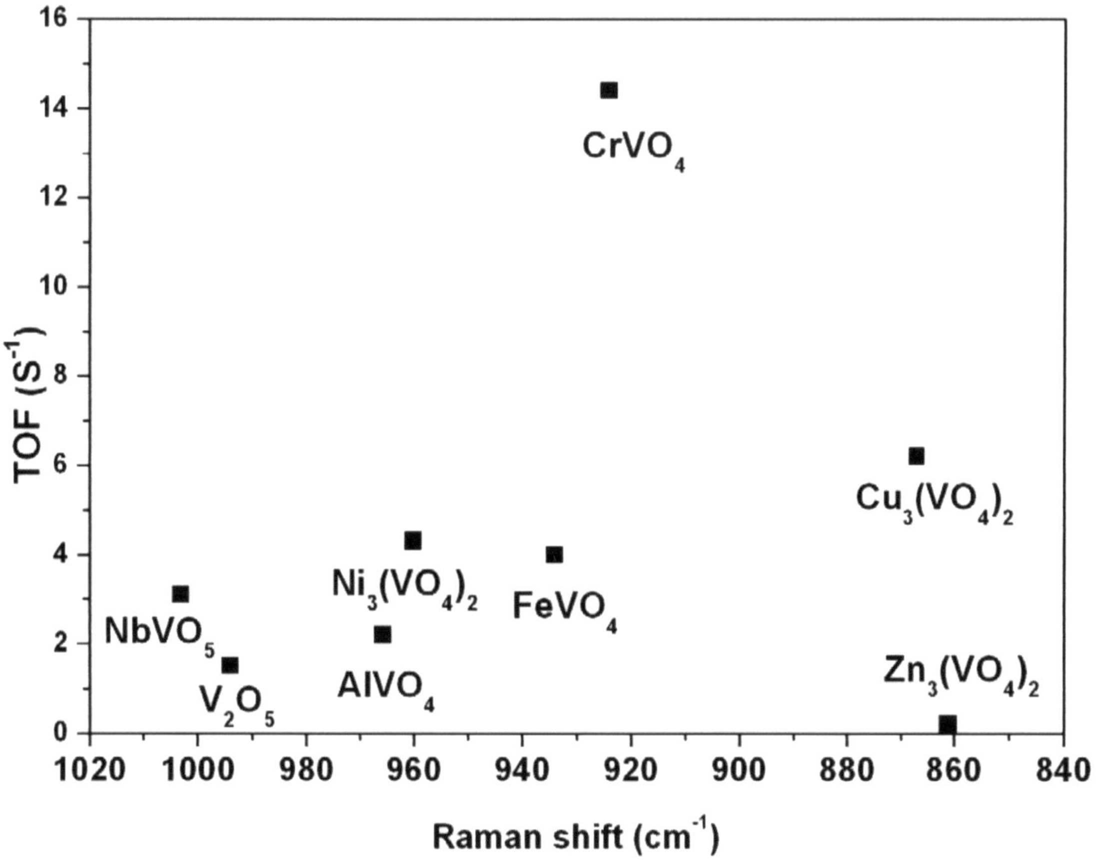

Figure 2. TOF vs. V=O Raman shift for methanol oxidation over bulk metal vanadate catalysts.

Conclusions

The reactivity of bulk and supported vanadium oxide catalysts are very similar in the presence of the same secondary oxide in the form of a support or a mixed oxide.

192c Redox Thermodynamics of Ceria and Ceria Mixed Oxides

Raymond J. Gorte, Department of Chemical and Biomolecular Engineering, University of Pennsylvania, Room 311A Towne Building, 220 South 33rd Street, Philadelphia, PA 19104

Ceria plays a critical role in three-way automotive catalysis because of its oxygen-storage capacitance. More recently, these same redox properties have been shown to make ceria a good promoter for steam-reforming, water-gas-shift, and hydrocarbon-oxidation catalysis. Scientifically, ceria is also interesting because its redox properties change when it is mixed with other oxides or prepared in a nanocrystalline form. In the automotive application for example, it is necessary to mix ceria with zirconia in order to have a material with sufficient thermal stability. This talk will outline some of the key applications for ceria in catalysis. The fundamental thermodynamic properties that allow these applications will then be discussed.

192d Advances In Computational Catalysis: Past, Present and Future

Matthew Neurock, Chemical Engineering, University of Virginia, 102 Engineers' Way, PO Box 400741, Charlottesville, VA 22904

This paper will review the past and recent advances in electronic structure methods and their application toward the elucidation and design of catalytic materials. In addition, we discuss perspectives on future developments.

193a An International Assessment of Research in Catalysis by Nanostructured Materials

Robert J. Davis, Chemical Engineering, University of Virginia, 102 Engineers' Way, Charlottesville, VA 22980

We discuss the findings from an international assessment of heterogeneous catalysis (commissioned by the World Technology Evaluation Center, National Science Foundation, Department of Energy, Air Force Office of Scientific Research and the Defense Threat Reduction Agency) that was recently completed.

193b History of Catalysis: An Industrial Perspective

John Armor, Global Catalysis, Allentown, PA 18195

An overview of some of the historical developments in industrial catalysis will be given.

193c The Complex Science of Catalysis and Catalytic Materials. A Historic and Forward Looking Perspective

Rutger A. van Santen, Inorganic Chemistry and Catalysis, Eindhoven University of Technology, Den Dolech 2 / P.O. Box 513, Eindhoven, Netherlands

This talk will present a brief historical review of major developments in catalysis, views on the future of catalysis resulting from an in depth study in the Netherlands, and the synthesis of bio-inspired nanomaterials.

193d Advances in Catalysis. A Historical Perspective

Burtron H. Davis, Center for Applied Energy Research, University of Kentucky, 2540 Research Park Drive, Lexington, KY 40511

Catalysis was practiced long before it was recognized. Berzelius is credited with the "discovery" of catalysis in the 1834-1836 period as well as introducing the controversial concept of "catalytic force." Mitscherlich had defined "contact action" at least two years before Berzelius introduced catalysis and had used as examples of the action essentially the same examples as Berzelius. Faraday, on the other hand, considered catalysis to fit into the general concepts of chemistry. Liebig introduced his "vibrational theory" in the 1840s and this view lasted until Ostwald, in the late 1880s, introduced his definition of catalysis that led to him recieving the 1909 Nobel Prize. Industrial catalysis began with the high pressure hydrogenation of nitrogen, and led to the Nobel Prize for both Haber and Bosch. Langmuir emphasized chemical adsorption, a view that dominated for many years, leading to the 1932 Nobel Prize. Physical adsorption was emphasized in the 1938 BET paper that is highly cited even today; this work was nominated for the Nobel Prize but never received the award. Beginning about 1930, petroleum refining was a driving force for catalytic advances, starting with Houdry's regenerative catalytic cracking proces, the introduction of the silica-alumina and then the crystalline zeolite catalysts as well as the introduction of the bi-functional naphtha reforming catalyst to produce high octane gasoline. The production of polymers using catalysis advanced rapidly, receiving recognition with the Nobel Prize awarded to Ziegler and Natta. The introduction and widespread use of surface science instrumentation led to major advances in the understanding of catalysis, recognized by the 2007 Nobel Prize being awarded to Ertl.

Future Directions in Surface Catalysis

Bruce C. Gates
Dept. of Chemical Engineering & Materials Science
University of California, Davis, CA 95616

Catalysis by solid surfaces is central to the technology of conversion of fuels; abatement of pollution; and synthesis of chemicals, pharmaceuticals, fertilizers, polymers, and nanomaterials. It seems almost inevitable that catalysis will increase markedly in importance in all of these applications. This presentation is meant to provide a summary of thoughts about where the field may be going on the basis of perceived needs for improved control of chemical conversions and recent advances in research that are advancing catalysis and related fields.

The largest-scale applications of catalysis are in fuel production and pollution abatement. With the energy and environmental challenges the world faces and the increasing realization of these challenges, it is evident that we need major innovations in both of these directions, and catalysis must be at the center.

Challenges in fuel conversion dictate the control of transformation of markedly heavier and dirtier fossil fuels, including heavy oil, oil-sands-derived liquids, coal-derived liquids, and shale-derived liquids; biomass also offers potentially significant sources of renewable fuels that must be assessed. The existing catalysts for heavy petroleum conversion provide a good starting point for the next generation of fossil fuel conversion processes, but much improved catalysts will be needed. It is essential that a detailed molecular understanding of the feedstocks be developed so that the various competing reactions can be elucidated and catalysts can be found that are tailored for specific feedstocks and feedstock fractions, to be converted in the presence of components that cause rapid deactivation of today's catalysts. The need for analytical chemistry of the complex feedstocks as part of the needed research in catalysis seems likely to set the research in these directions apart from earlier research in fossil fuel conversion.

Biomass conversion catalysis is in its infancy, and it is difficult to assess its prospects for economic application. The high oxygen content of biomass leads to new challenges in catalysis, as the reaction pathways and catalyst deactivation phenomena are qualitatively different from those of almost all existing processes and will seemingly require classes of catalysts that are essentially different from today's.

Conversion of effluents from motor vehicles and power plants has become extremely effective and provides a strong demonstration of the social value of catalysis, but innovations are needed to facilitate catalytic solutions of the challenge of converting CO_2 formed in fossil fuel conversion and converting water into H_2 and O_2. The needs suggest opportunities for photocatalysis and processes emulating photosynthesis, and the prospects of high-temperature catalysis are also tantalizing, but largely unexplored.

Extensive research will be needed for landmark advances in these directions. The opportunities for major advances in research are excellent, facilitated by continuing rapid advances in experimental methods of surface characterization and theory and computation, bolstered by novel methods of materials synthesis. Exploitation of the improving and emerging scientific tools, at least in the U.S., will require a funding base that is far stronger than today's.

Understanding of catalysis on surfaces is challenging because surfaces of most catalysts are dauntingly complex and nonuniform and work at high temperatures and often high pressures—and because the catalytically active species are small and dispersed within solids and difficult to characterize. Most of the reported experimental results characterizing catalysts represent them under conditions far removed from those of catalysis—so that the structures investigated are often not entirely relevant to catalysis.

The techniques for investigation of catalysts in reactive atmospheres continue to improve markedly and rapidly—the application of spectroscopic and even microscopic techniques to working catalysts is a major trend in recent research and is likely to continue strongly.

Fundamental understanding of solid catalysts is increasingly expected to emerge from investigations of catalysts that are structurally uniform—the successors of single crystals of metal, such as well-made zeolites—so that the catalytic sites constitute uniform arrays. This class of catalytic material can be extended by anchoring analogues of homogeneous catalysts ("molecular" species) onto solids with uniform surfaces, so that the characterization by spectroscopy can be incisive and theory can be applied with a rigor approaching that in molecular chemistry.

What's more, it is becoming possible to investigate individual catalytic species (nanoparticles) in the presence of many other species, and we may anticipate further progress in this direction. Such results will set a foundation for truly fundamental understanding of how individual catalytic species work, even in highly complex catalytic materials. Some day, we envision single-molecule spectroscopy for characterization of molecules reacting on individual catalyst structures. Ultimately such results will allow elucidation of truly fundamental structure-performance relationships in catalysis, so that the results can be integrated over whole, complex catalysts.

Black powder manufacture

Submitted by Robert A Howard

As part of the session on particles this paper concerns a thousand year old product named variously as powder, black powder or gun powder (The black powder designation occurring in the 19th century when other gun powders were created.) Black powder is a physically combined mixture of four ingredients. These are classically potassium nitrate, sulphur, charcoal and water. Because a powder is black in color, it is not necessarily the chemical black powder. Other powders, as we know them today, are single of double based nitrated hydrocarbons.

The largest and most sophisticated black powder plant ever built anytime, anywhere, is about 30 miles south of Philadelphia just outside Wilmington Delaware. It was the basis of the Dupont Company, which, now a little over two hundred years after the founding has emerged as one of the world's largest chemical companies. Dupont sold the last of its black powder plants in 1972 to Gerhart Owen which is now GOEX. Part of the first Dupont plant survives as the Hagley museum which has associated with it a library containing the best historical collection of explosive related materials in existence.

Although the product has been known for about a thousand years, the dates being somewhat disputed, accurate knowledge of its burning is only about one hundred and fifty years old. Black powder burns or technically deflagrates. (as opposed to detonates which is a considerably faster reaction) Since black powder has various applications from blasting, to shooting pistols to firing cannons its utility is a factor of controlling the deflagration. To put it simply, gas generation needs to be factored by time to create the ideal burn. In blasting one needs to heave or shatter. In shooting the time of the burn is factored by the diameter of the barrel and the length. To get a continuous push (called a progressive burn) the burn needs to be complete while the projectile is in the barrel. Since barrels vary in length from about an inch for a derringer to several feet for a piece of artillery the needs are considerably different.

There are two ways to control the burn. The first and oldest was to vary the proportion of the ingredients. The other and more sophisticated is by varying the physical means of creating the powder grains. Grain is a rather loose term which encompasses the tiny particles of the early serpentine powder to the large pressed pieces of prismatic or pellet powders. There are two basic ways to control the mechanical process. The first is to simply vary the size of the grain, and the second is length of time of each manufacturing process or the nature of it. Time is tricky because after a point continuing a mechanical process has no benefit, but carrying it out to the ultimate benefit is not economical in some cases (as in cheap blasting powders).

Powder 2

To continue, we will consider the four ingredients first, and then we will discuss the evolution of the mechanical processes and the ramifications of each to the burning reaction.

The major ingredient is potassium nitrate which is about three quarters of the mixture. Its purity is critical. While other nitrates have been substituted such as sodium nitrate for cheap blasting powders and ammonium nitrate, potassium nitrate is the material of choice. It is the least hygroscopic (does not as readily absorb atmospheric moisture). Because the process of extracting nitrate historically required the leaching and refining of waste materials, the early powders may well have had some calcium or sodium nitrate. This was solved by running a liquor of nitrate solution through potash. Purity was important for two reasons. The first is that anything else does not properly participate in the burn and the second, materials such as chlorides remain in the solid residue and degrade the weapons. Fortunately the process for making potassium nitrate today is not the long arduous and stinky process of old. However, issues of fertilizer grade nitrates verses reagent grade nitrates occasionally arise. Occasionally the fertilizers may contain chlorides in small amounts. The ability to artificially produce potassium nitrate is about 150 years old, but there was immediate switch to it for about a half century.

The second ingredient is sulfur or brimstone. Getting pure sulfur is relatively easy with a sublimation reaction. Simply one drives it to a gaseous state and then condenses the gas. Since sulfur is an element, it is relatively simple. It combusts with extremely noxious odors and causes great distress in the gastric tracts it ingested.

The most complex issue is the charcoal. The structure of the charcoal is determined by the wood from which it is distilled. Usually it is about 70% carbon, and the more porous the maze, the better. Classically willow, alder and hazel have been the woods of choice while the hard woods prove considerably less satisfactory, Small amounts of creosote in the charcoal produce a superior *moist* burning powder. The trace amount of ash kept the powder slightly alkaline making stable grains. The carbon takes place in the burning reaction but the reaction is a bit complicated. Carbon dioxide is about 49 %, carbon monoxide is about twelve and a half percent of the gaseous products while solid products include potassium carbonate(61%) , potassium thiocyanate (.2%) / pure carbon (.8%) and ammonium carbonate (.8%). Since wood is a hydrocarbon, there is some methane, hydrogen sulfide and pure hydrogen released as gases and the previously mentioned ammonium carbonate solid residue. Originally the charcoal was produced by controlled burning of mounds of wood, usually taking a couple weeks. Around the turn of the 19th century the process of distilling the wood in large iron vessels was introduced to powder making, and by the turn of the 20th century the use of superheated steam was in use to substitute for the fires under the vessels. Dupont also collected the tars and pyrolignious acids in long collection tubes for resale and internal use. The distillation was more accurate and did not have the hazard to the burners of becoming part of the burn by accidentally falling into the smoldering mound.

Water is the final ingredient – usually neglected. It is by weight about ½ of 1 percent of the final product and has a dual role of binder producing a stable grain, and reduces the explosion hazard during manufacture. Water purity was recognized as a factor in the 19th century by W.W. Greerner is his famous scholarly tract The Gun and Its development (London 1835 ...first edition of about 9). More recent work at Aberdeen Proving ground and at the decontamition of the California Powder Works site has pin pointed and used the presence of micro-organisms as the cause of some powder degradation.

Another material associated with black powder is graphite. This is usually added in the glazing process. A myth in powder making is that graphite coating of the grains makes the grains less hygroscopic while in fact it makes it look better and pour better. It was initially (introduced in the early 19th century) the favorite of the African trade in which a 6 ¼ pound keg of powder was a trading unit for a slave. Not all powders were coated with graphite and it does not appear to contribute to the burning reaction at all.

Although varying the composition of the powder, most particularly the amount of nitrate, appears in the early literature as the means of controlling the burning reaction, the means of mechanically producing the powder is the real key to control. The process evolved as new procedures were invented and modified.

The early powders were called serpentine powder. The ingredients were measured out, pounded together in a mortar and pestle, and then used. This was usually done at the site of use since serpentine powder did not transport well. Serpentine powder was generally fairly weak powder because the particles were tiny. While generally the smaller grains tend to burn faster there comes a point when the particles are so small that they retard the burn rate. This is because with the larger particles the fire does not have to ignite each grain sequentially but may pass through the space between grains. With serpentine powder or powders of a granulation above 4F the particles physically block the passage of the fire and each needs ignited before the next can combust.

Powder 4

The basic process of combining the ingredients is called incorporation. Originally it was done with a mortar and pestle. As early as 1607 there is evidence that an edge mill (roll mill) was known to be used. Other means were tried such as heating the nitrate and pouring in the other ingredients , mixing with shear rolls, and using a jet mill. The first tended to blow up frequently, the second made some rather wretched blasting powder and the jet mill proved unsatisfactory. Interestingly the jet mill was a Scandinavian invention used in a fire proof powder plant which blew up and was not rebuilt. The U.S. government then built an experimental plant in Indiana using the jet mill which could only make fuse powder and was subsequently mothballed.

Essentially only the mortar and pestle and later the edge mill – using two cast iron wheels over a cast iron bed plate - were satisfactory. This is because mechanically these two means shear an air surface off the ingredient's particles, placing them in direct contact which aids combustion. The length of time that the process takes place is also critical. When the large 10 ton cast iron wheels were used , it was found that four hours was optimum. Less time and the powder was weaker because the work was incomplete; more time would produce no additional benefit as the work was complete. Also critical was the amount of water added. Too little water and the mills would "dust" and tend to blow up. Too much water and the wheels would form "clinkers". These solid chunks would cause a wheel to rise up, driving the other wheel through the bed of powder and hitting the cast iron bed plate. Often this would cause an explosion also.

The powder was then dried to remove most of the water which had been previously added. This process evolved. It was originally done in ovens resembling the beehive ovens used in cooking. A fire was built in the oven and then removed when the oven was hot. The powder was then put in the oven and dried by the heat. Powder was also dried on large outdoor tables. Later gloom stoves were invented and rooks were built with stoves. The room was heated by the stove and then the fire was put out. The powder was then placed in trays in the heater room to dry. Finally the rooms were heated from the outside by either fire places or steam.

In the later nineteenth century the whole drying process changed when the glazing process (discussed later) and the drying process were combined in large barrels. At first the drying was done by "frictioning" in which the tumbling of the grains in the barrel produced heat. When the barrel reached the 12 o'clock position the door or bung opened and the moisture escaped. Later hollow trunnions (bearings) were introduced and air or steam could be shot into the barrel. The air dried the powder and a short shot of steam helped the graphite to stick to the grains.

Historically the first quantum leap in powder making occurred around 1440. This was graining or corning. Moist powder was forced through sieves producing larger grains of powder. This was a huge improvement because the power of the powder increased as the burn rate increased. With the grains the fire did not have to ignite tiny particles sequentially but could pass between grains igniting grains further down the charge. By varying the size of the grain the burn rate could be controlled with the larger grains being slower to combust than the smaller grains. Also equally as important, powder grains were transportable. Obviously this transformed warfare – ships became floating gun platforms and while sieges continued other forms of land battles were possible. The era of the knight in armor faded (although the long bow was also instrumental).

After the powder press was common in the early 19th century the graining process changed. Instead of forcing powder through sieves, cakes of powder were broken up by roller grinders and the grains sorted by size in sieved reel type machine called bolters. The roller grinder was invented by William Congreve who also invented the rocket alluded to in the national anthem… rockets red glare. Controlling the size of the grain controls the ease of combustion and the speed of combustion.

Beginning in the mid-nineteenth century super sized chunks of powders were produced for use in large artillery and blasting. This was mostly the work of Rodman aided by Lammot duPont .Prismatic powders which were hexagon pressed pieces of powder with holes in the center to aid ignition used in artillery. Pellet powders, used in blasting, were cylindrical pressed chunks of powder. In both cases the objective was to slow the generation of gases during combustion.

The next procedure added the process was glazing (about 1660).This takes place after the graining. Not all powders were glazed. Powder sold as rough powder(mostly blasting) and molded powders (prismatic and pellet powders) were not glazed. Essentially the grained powder is tumbled in barrels. Glazing simulated the rigors of transportation. The grains were smoothed and rounded. The inferior grains were destroyed and the resulting dust was removed and recycled. Glazed powders were best for applications some distance from the point of manufacture. This was especially critical in nautical applications because if the powder was shaken it might degrade to smaller grains and the burning characteristics change.

Powder 6

The big change in the glazing process occurred in the later part of the nineteenth century when small barrels were replaced with large barrels and drying was combined with glazing. It was discovered that glazing changed the nature of the grains. If the glazing were done for some time, the nitrate would move to the surface of the grain. This seems to aid the combustion. The industry term is called a *hard glaze*.

The final mechanical process to be invented was pressing. It was started in England in the mid eighteenth century but seems not to be common until the beginning of the nineteenth century. For example E.I. DuPont did not know of it when he worked with Louvosier, but knew of it when he returned to America in the very beginning of the nineteenth century. Pressing occurs after incorporation. It makes the powder more dense making it be able to produce more gas per unit volume. The process underwent only one change which began in the middle of the nineteenth century. A hand powered mechanical press was replaced by a powered press. The earlier presses were screw presses and the later presses were hydraulic There is a press box in which the powder is pressed. To make manageable sized pressings there are separator plated held initially in place by finger boards. These are a comb affair with teeth about two inched apart. When the powder is shoveled into the press box with the separator plates, the finger boards are removed and the powder is compacted. The plate pressures used to be about 3000 pounds per square inch, but current practice is a bit less. After pressing the powder is in the form of slabs resembling slate.

These slabs are initially broken up into chunks by a grinder which consists of two bronze rollers with teeth. After it is made into chunks it is subsequently made into grains by roller grinders. These have between three and four sets of opposed rollers.. Some powders (prismatic and pellet) are pressed into their final form and not further reduced.

The big effect the pressing process had was to reduce the volume of powder to get the same effect.

A final step was added around the beginning of the nineteenth century. The French called it the *revolutionary* process which became *pulverizing*. Initially the three ingredients were combined in rotating barrels with heavy metal balls before incorporating. These mills were called dust mills and had a bad habit of disappearing. After the Civil war the American practice was to combine the charcoal and the sulfur only in the barrels. Without the nitrate the hazard of explosion was significantly reduced.

When the powder was complete it was sent to a pack house where the dust created by the glazing was removed by a bolter – to be recycled -and the result put in kegs, cans or papers.

Dupont did experiment with a cardboard type keg called a pulp keg around the turn of the 20th century.. These were less than satisfactory having a tendency to collapse when stacked, and when the "fire proof" building in which they were made burned down , the effort was suspended.

How dangerous is making black powder? In the nineteenth century this was relative. Being a farmer (dying of an accident or tetanus) , railroad brakeman (being crushed before the invention of the Westinghouse air brakes) or hatter (dying of mercury poisoning) was far more dangerous. The most likely explosion to occur was in the incorporation process. Usually these happened at start up or shut down. Since the rolling mills were run by remote control, the loss of life was minimal. Also since the powder was not confined the explosion was not particularly violent. The other mills were considerably more dangerous because there was a larger quantity of more refined powder in a confined area. Since powder was made in a series of separate buildings, the risks were confined, however major losses did occur. The worst explosion at Hagley occurred in 1890. Over one hundred TONS of powder went up. The shock wave hit New York City. Trees a quarter mile from ground zero were blown over.

There were architectural considerations in making powder mills. These usually consisted of making buildings with heavy three walls and a light fourth wall and roof to channel an explosion, or building all walls and roof very light to minimize the amount of debris flying about. Later in the nineteenth century barriers were erected with wooden walls filled with dirt. These were called Rapauno barriers and were designed to stop flying debris. As with auto wrecks, many of the explosions were alcohol relater.

Today there is one black powder maker in the United States and several in the world. Interestingly the general quality of powder and the knowledge of its making has declined world wide. There is no plant in the world as sophisticated as Hagley was in the early 1900's. It could make the blasting powders, the sporting powders, the military powders, the prismatic powders, the pellet powders and the semi-smokeless powders. Maybe this suggests that progress is not linear but a series of undulations.

224b History and Development of Salt Production Processes and Crystals

Roma Gimelfarb, Vacuum Crystallization Technology, Morton Salt, a Rohm and Haas Company, 123 North Wacker Drive, Chicago, IL 60606-1743

This presentation will provide an historical perspective on the role of salt throughout history and transformations in the ways salt is utilized and produced. It will discuss how cultural and consumer demands affected changes in the quality of salt and morphology of crystals. It will review the development of salt production technology through the ages and future production processses.

224c History of Solids Processing at the Dow Chemical Company

Karl Jacob, The Dow Chemical Company, Dow Chemical Company, 1319 Bldg., Midland, MI 48667

The Dow Chemical Company has had a rich history in particle technology from the early days of aspirin, MethocelTM, polystyrene and StyrofoamTM to the myriad of present day products including specialty polymers, agricultural chemicals, and water soluble polymers. This paper will look back at the development of particle-based intermediates and products and how they shaped the company's direction to its present position today. Key in the development of these products has been the people who invented and developed manufacturing processes for them. That traditional continues with the existence of the Solids Processing Disciplines and Technology Centers in place at Dow today. Lastly, the paper will look at the challenges which lie ahead in solids processing into the 21st century.

Methocel and Styrofoam are trademarks of The Dow Chemical Company

224d Progress toward First-Principles Design of Particle Processes at Merck

James N. Michaels, Pharmaceutical Materials Laboratory, Merck & Co Inc., PO Box 4, WP78A-31, West Point, PA 19486

Realized examples of the desired "design-then-build" paradigm will be reviewed followed by a discussion of knowledge gaps that impede progress.

224e History of Aerosol Manufacture of Materials

Sotiris E. Pratsinis, Particle Technology Laboratory, Institute of Process Engineering, Department of Mechanical and Process Engineering, ETH Zurich, Sonneggstrasse 3, Zurich, CH-8092, Switzerland

The historic development of commercially available, aerosol-made particles is reviewed. Emphasis is placed on flame-made carbon black, TiO_2 and SiO_2 that dominate the aerosol-made material market by volume and value followed by particles made in hot-wall reactors (ZnO, Ni, WC). Despite the initial empiricism associated in creating such technologies, today there is good understanding of the controlling mechanisms based firmly on aerosol science and engineering that focuses on the *high temperature residence time* of these particles. In fact, this understanding guides the rational design and construction on new units of the above commodities outside their traditional base in N. America and Europe. Furthermore and, most importantly, it propels aerosol technology as an enabling one to develop new processes and products for catalysts, sensors, biomaterials, electroceramics, photonics and others. On the other hand, economic manufacturing of such materials brings new challenges to aerosol science as, for example, with the dynamics of fractal-like aerosols flowing at high concentrations (1), health effects and the synthesis of multicomponent, layered particles (2).

1. M.C. Heine, S.E. Pratsinis, "Agglomerate TiO_2 Aerosol Dynamics at High Concentration", *Part. Part. Syst. Charact.* **24**, 56-65 (2007).

2. R. Strobel, S.E. Pratsinis, “Flame aerosol synthesis of smart nanostructured materials”, *J. Mater. Chem.*, **17**, 4743 - 4756 (2007).

224f The Development and Commercialization of the DuPont Chloride Process for the Manufacture of Pigments Based on Titanium Dioxide

Austin H. Reid Jr., DuPont Titanium Technologies, 7685 Kiln-DeLisle Road, Pass Christian, MS 39571

The DuPont Chloride Process for the manufacture of pigments based on titanium dioxide was originally developed in the 1940's and 1950's, with full commercialization in the 1950's. The process was a radical departure from the previously-established sulfate-based manufacturing route. Its development and commercialization required DuPont scientists and engineers to overcome a myriad of problems in materials of construction, process control, aerosol science and other arenas. The process has been continually improved in the intervening fifty years and is the underlying basis for the production of all DuPont titanium dioxide pigments.

This presentation will provide a historical perspective on the development of this revolutionary process, its commercialization and proliferation, and a discussion of some important aspects of particle science and engineering inherent to its success.

237a The Interplay Between Simulations and Thernodynamic Models

Berend Smit, Deptartment of Chemical Engineering, University of California at Berkeley, 20, Gilman Hall, Berkeley, CA 94720-1462

The Interplay Between Simulations and Thermodynamic Models

237b Perspectives and Prospects for Thermodynamic Modeling

Walter G. Chapman, Chemical and Biomolecular Engineering Department, MS-362, Rice University, 6100 Main St., Houston, TX 77005

Over the past twenty years, significant advances in statistical mechanics have provided engineers with accurate equations of state for bulk phase behavior of mixtures containing associating, polar, and polymeric components. More recently, developments in statistical mechanics are providing scientists and engineers with valuable tools to correlate and predict bulk and interfacial phase behavior and microstructure of complex systems containing block copolymers and surfactants. Insights will be provided in the development of these theories. Opportunities for applications and extension of these models also will be discussed along with future challenges.

237c Predictive Thermochemistry from Quantum Chemistry: Current State and Perspectives

Andreas Klamt, COSMOlogic GmbH&CoKG, Burscheider Str. 515, Leverkusen, 51381, Germany

In this talk the current state of quantum thermochemistry will be analyzed with respect to its predictive potential for molecules of typical industrial engineering size. It will be shown that the recipes developed for small molecules can hardly be extended to molecular sizes of industrial interest, and that they even turn out to become partly invalid for larger molecules. Some viable protocols for larger molecules will be mentioned together with crude estimates of their accuracies and computational costs. Since most quantum thermochemistry methods so far are limited to the gasphase, while in industrial applications mostly fluid phases are relevant, some aspects of solvation for thermochemistry will be discussed.

Web Page: www.cosmologic.de

237d Simulation of Electrolytes: Model Development and Industrial Practice

Andre Anderko, OLI Systems, Inc., Morris Plains, NJ 07950

Simulation of Electrolytes: Model Development and Industrial Practice

Recent progress in modeling thermophysical properties of electrolyte systems is reviewed. In particular, attention is focused on models for mixed-solvent electrolyte solutions and equations of state for high-temperature and supercritical electrolyte systems. The models are analyzed with respect to their capability of computing thermodynamic and transport properties in wide ranges of conditions and composition (e.g., for aqueous or mixed-solvent, dilute or concentrated solutions). A methodology is discussed to leverage the insights gained from electrolyte thermodynamics to predict corrosion of engineering metals. Further, applications of electrolyte models are reviewed for topics including the design of nuclear waste repositories, hydrometallurgy, crystallization and synthesis of inorganic materials in hydrothermal media.

"Energy Applications: Impact of Data and Models"

Energy Applications refers in this particular case to the wide application of equations of state upstream in the Production of Oil and Gas. The petroleum business is run on the basis of gas volumes at STP and oil volumes in stock tank barrels. The impact of data and models refers to the use of equations of state as representation of reservoir fluid to provide fluid properties as whatever conditions of pressure, temperature, composition and shear are involved in reservoir and production operations in the petroleum industry.

Equations which can relate the volumes at sales conditions to those in the reservoir have value in aiding a correct assessment risk in reservoir development. These same equations are used to forecast well performance test and reservoir evaluation, as well as developing process engineering that optimize the resource. These same equations are used in property generation for transient multiphase flow simulators to determine the operational envelope for deepwater production facilities. These tools are now used in live real time simulations models that aid field operations in making real time decisions on potential operational problems.

Similar fluid models constructed from data and equations of state have been used in enhanced oil recovery for many years especially where mass transfer effects are important in the reservoir process. An example of such can be found in miscible and immiscible gas recovery. These tools are also used in transportation and process simulation in LNG process design as well.

This talk describes the successes and short comings of such equations as presently used and the needs in this area for the future of the oil and gas business.

Equations of state classically relate the thermodynamic variables molar volume, pressure, and temperature. These equations have been effectively used for complex mixtures with relatively simple compositional mixture rules for a variety of applications during designing reservoir processes and facilities from the first discovery through the production of an asset to completion. The purpose of all EOS applications is the successful development of a model of a given process fluid which can effective relate the properties of the fluids to changes in pressure, temperature, volume, and composition during a process of interest.

The strength of these equations is that they may now be used for evaluation of the quality of measured property studies performed to define reservoir fluids as well as a tool for simulation of properties consistent with a measured basis set for other conditions in the process. Typically in the past the quality measurements were restricted to material balance and "K" value correlations such as Hoffman, Crump, and Holcott[1]. The contractor community also realizes this and typically "smoothes" the reported values so that it is difficult to asses errors in experimental data.

[1]Hoffman, A.E., Crump, J.S., and Hocott, C.R., "Equilibrium Constants for a Gas Condensate System," Trans. AIME (1953) **198**, 1-10

Another example would be the series of separator tests used in the past to determine the optimum conditions for separator of gas and oil in a field development. The advent of cubic equations and application of these with personal computers allows ready optimization of conditions based on a single well done separator test.

In some situations, such as lean gas systems, where the determination of dew point is a key variable in reservoir simulation, many times the simulated dew point values may better estimated with a well determined composition and EOS compared with direct measurement.

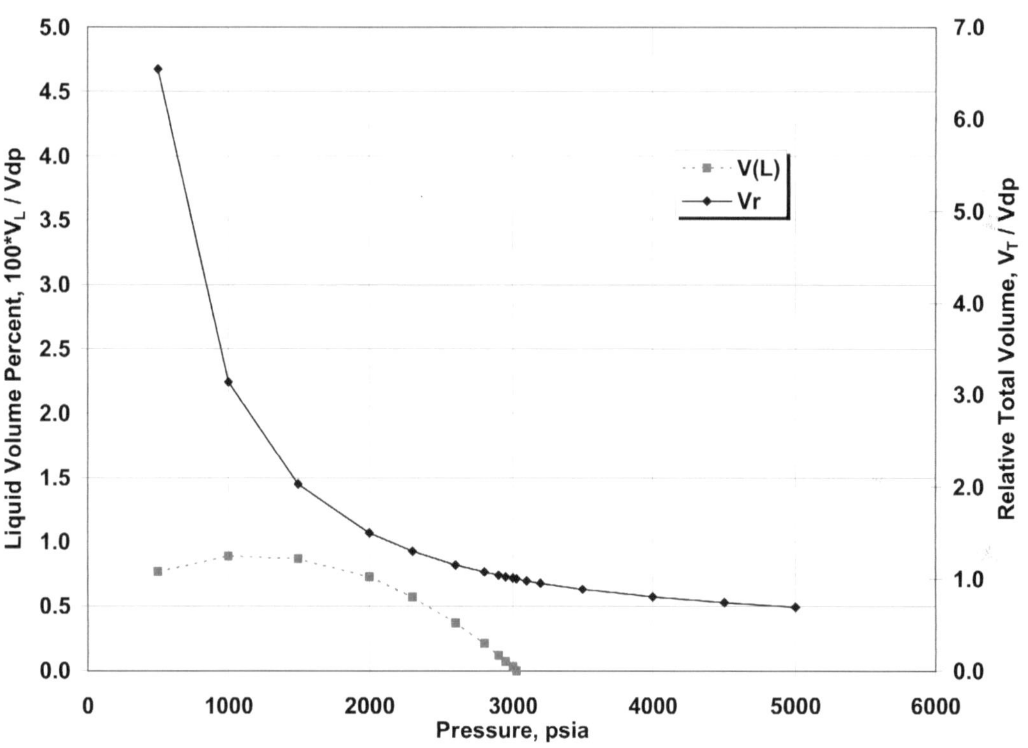

Figure 1 Simulation of Dew Point Pressure

There are however suggestions that the current cubic equations are insufficient for present and future reservoir assets.

There are fundamental problems that still exist with the temperature and pressure dependence of the predicted densities.

The temperature dependence continues to be a problem as equations of state are used in commercial applications involving flow lines and wellbores for example. In these situations the temperature may range from more than 300 °F in the reservoir to 40 °F in the subsea pipelines. The inability of the equations to predict the correct thermal expansion of fluid mixtures leads to problems in both the phase behavior as well as the simple density calculations. This also plays a role in the ability of the EOS to correctly represent interfacial tension and viscosity since calculations of both these variables requires reasonable density estimates. Figure 2 below shows the difficult with the dependence in the fluid density as the $Z = PV_m/RT$ as a function of both temperature and pressure with either the traditional Soave Redlich Kwong[2] or the Peng Robinson[3] equations of state.

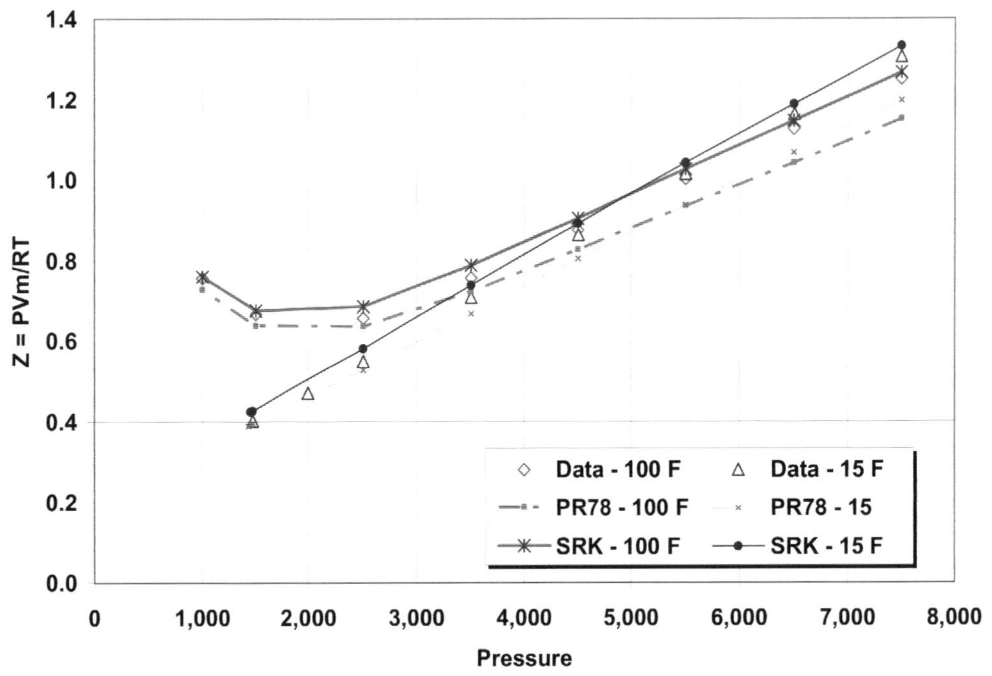

Figure 2: Variation of Z values with Pressure and Temperature

The temperature dependence in the cubic equation of state aside from the explicit temperature dependence from RT is from the a(T) term which depends on reduced temperature, T/T_c and Pitzer's acentric factor $\omega = -\log(P_v^* / P_c) - 1$ where P_v is the vapor pressure at a reduce temperature $T/T_c = 0.7$. The temperature dependence of the pair potential is then given by

$$a(T) = \left(1 + \kappa \cdot (1 - T_r^{1/2}\right), \text{ where } \kappa = 0.37464 + 1.54226 \cdot \omega - 0.26992 \cdot \omega^2$$

The current correlations provide a variety of interesting and amazing results for acentric factors including some that show the acentric factor going to zero for large molecules while others give values as high as 2 to 3 for the same molecular weights. Our experience is that a maximum value of 1.2 to 1.3 is most likely and correlates well with the temperature dependence of the bubble point pressure for crude oil systems.

[2]Soave, G., "Equilibrium Constants From a Modified Redlich-Kwong Equation of State," Chem. Eng. Sci., (1972) **27**, No. 6, 1197-1203.
[3]Peng, D.Y., and Robinson, D.B., "A New Two-Constant Equation of State," *I. &E.C. Fundamentals* (1976) **15**, No. 1, 59-64.

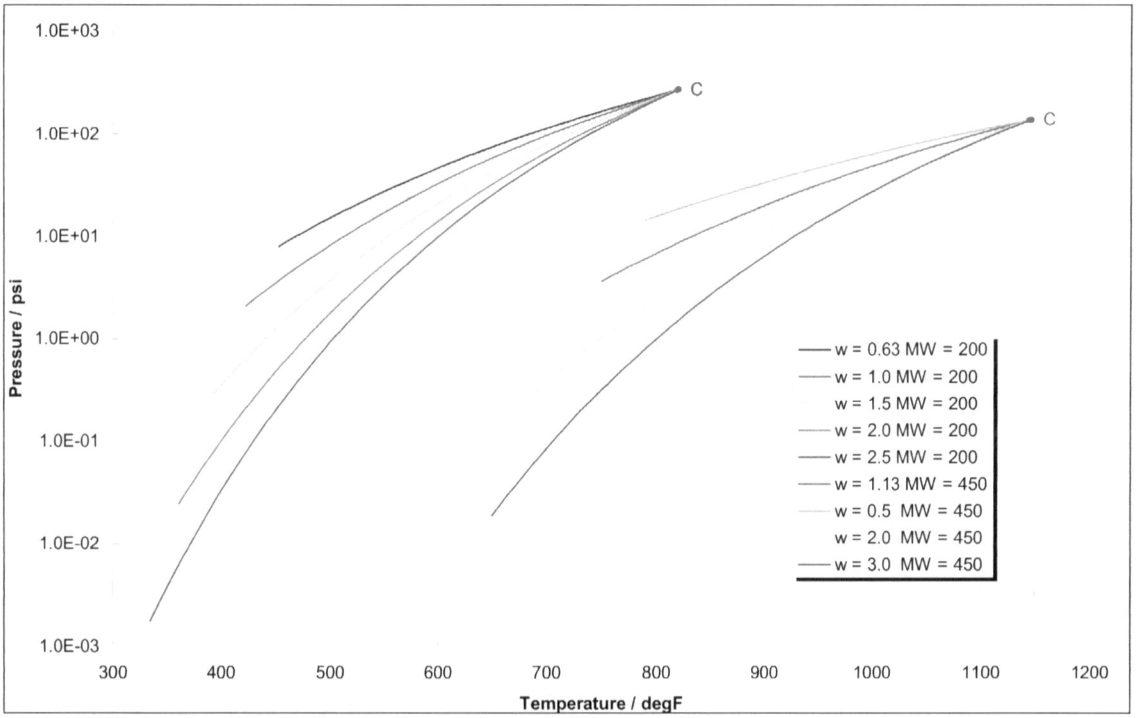

Figure 3 Variation of Vapor Pressure with Acentric Factor.

The following vapor pressure curves were made for pseudo components of differing molecular weights and showing the variation due to acentric factor variation. The values of 2 or 3 show unreasonable vapor pressures for hydrocarbons of effectively below 0.1 psi for any reasonable reservoir temperature (T_{res}(max) < 400 °F)

There are also problems with the pressure dependence as well. The typical equations of state are cubic in volume. This implies that the compressibility can be at most a linear function of pressure The problem to be illustrated as reservoir simulators must relate fluid properties in the single phase region above saturation by correcting bubble point data with compression terms. The newer reservoir fluids have high pressure reservoirs (~ 20,000 psia) while also having bubble point pressures of less than 3,000 psi thus requiring accurate compression calculations over long ranges of pressure.

An example of this is shown in the figure below where the ability to forecast the reservoir volumes with typical simple compressibility where the compression can vary only linearly with pressure fall short of matching the reservoir fluid data. The volumes are expressed as relative volume to the bubble point condition in the particular experiment.

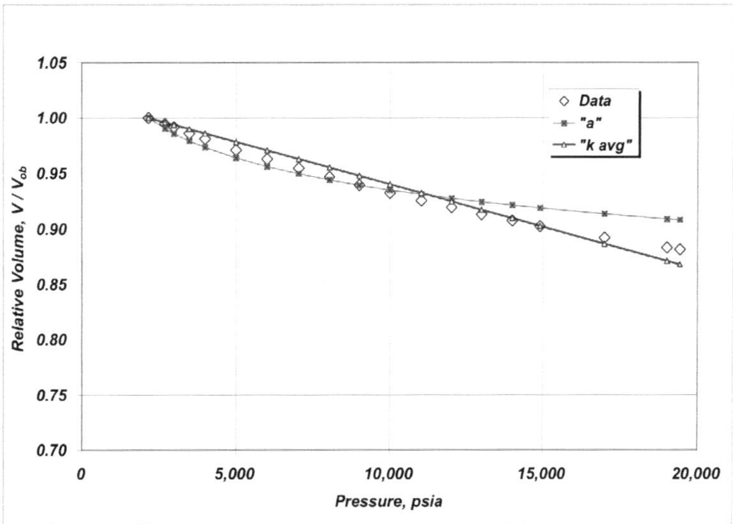

Figure 4: Compression corrections for Reservoir Simulation

Figure 4 shows two attempts at applying simple constant compressibility correction to estimate formation volume from the bubble point volume. The problem is not resolved having the compressibility as simple function of pressure.

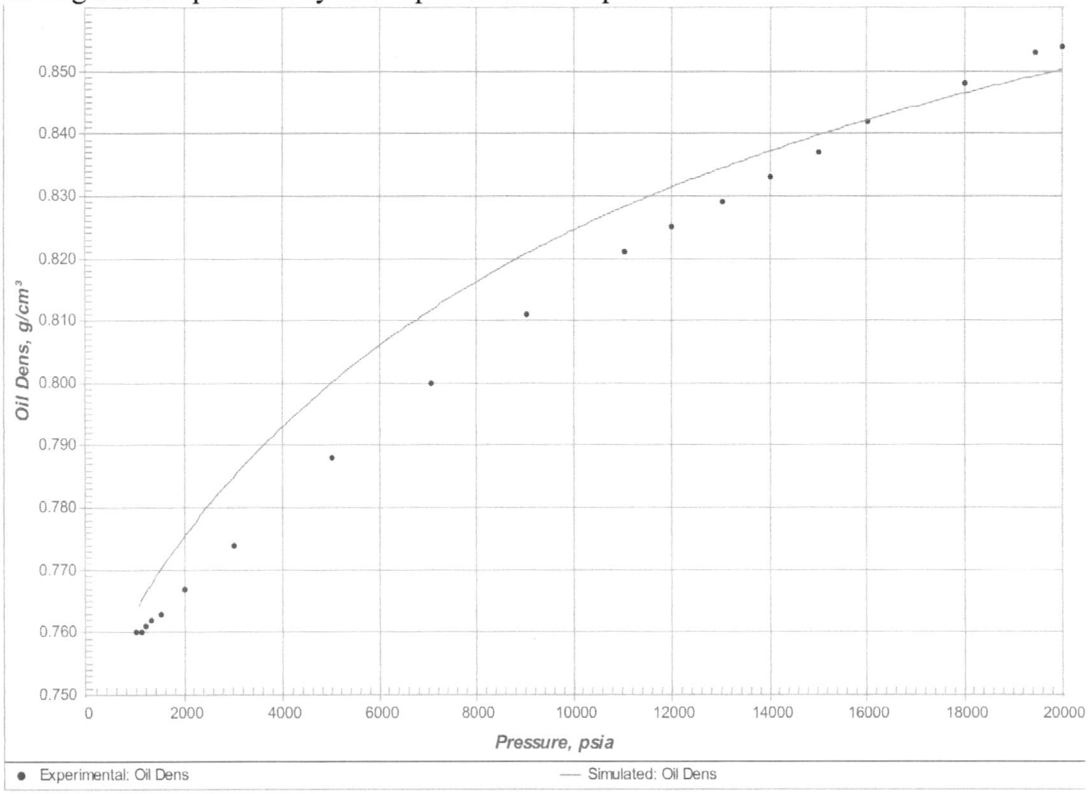

Figure 5: Typical fit of experimental density with equation of state.

Figure 5 shows a cubic equation of state SRK with volume shift density predictions compared to the experimental data. The incorrect dependence of density on pressure is shown as the curvature from the equation of state relative to the experimental data.

Figure 6 and 7 compare data from the NIST Chemistry Web Book and P.W. Bridgman's "The Physics of High Pressure" against the prominent equation of state tools.

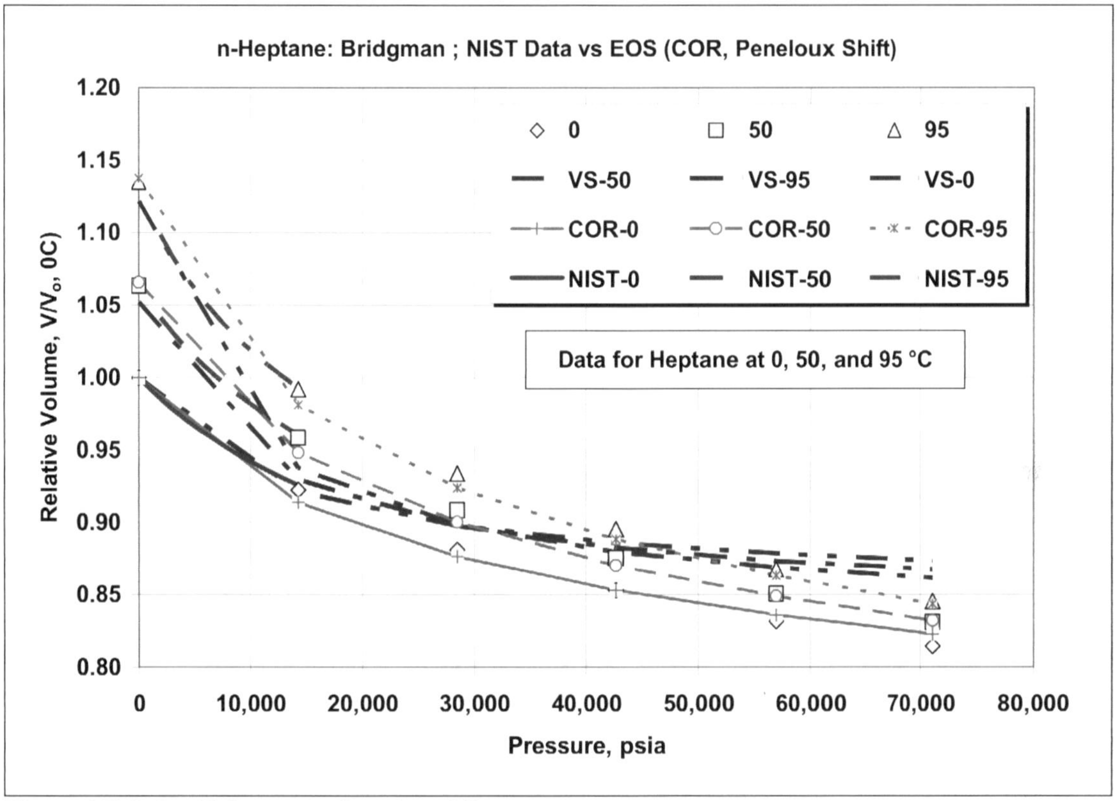

Figure 6 Relative Volume as a function of Pressure

COR is Chain of Rotators from M.C.H. Chien and M.R. Monroy, "Two New Density Correlations," SPE 15676 , SPE ATCE 1986. (Chien, M, C. H,, Greenkorn,R. A., and Chao, K. C, "Chain-of-Rotators Equation Of State", AIChE J. (July 1983) 29, 560-571).

VS refers to volume shift parameters in the cubic equations of state.
Peneloux, A., Rauzy, E., and Freze, R., "A Consistent Correction for Redlich-Kwong-Soave Volumes," Fluid Phase Equilibria (1982)8,7-23. and for Peng Robinson equations of state, Jhaveri, B, S., "Three-Parameter Modification of the Peng-Robinson Equation of State to Improve Volumetric Predictions," SPE Paper 13118 presented at the 59th Annual Technical Conference and Exhibition of the SPE of AIME in Houston,TX, September16-19, 1984

Figure 7 compares one of the newer equations of state, PC-SAFT with P.W. Bridgman data for n-heptane and benzene.

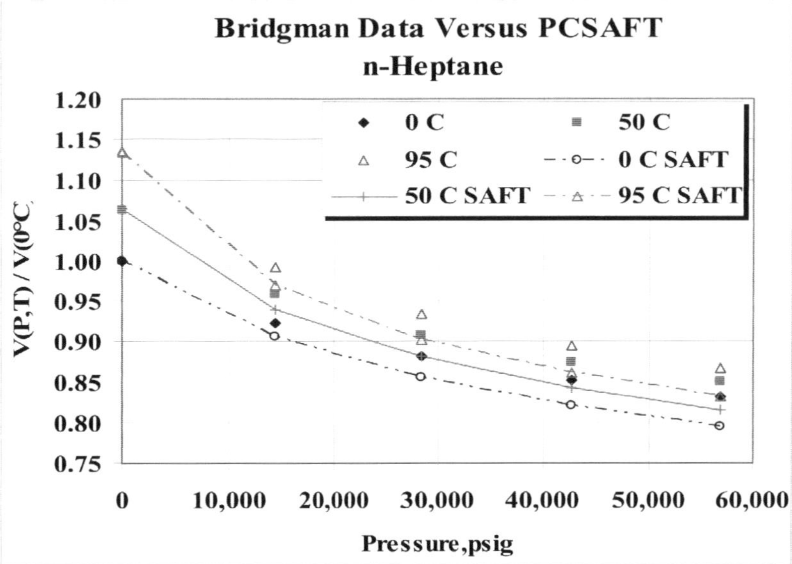

Figure 7 SAFT versus Bridgman Data for Heptane

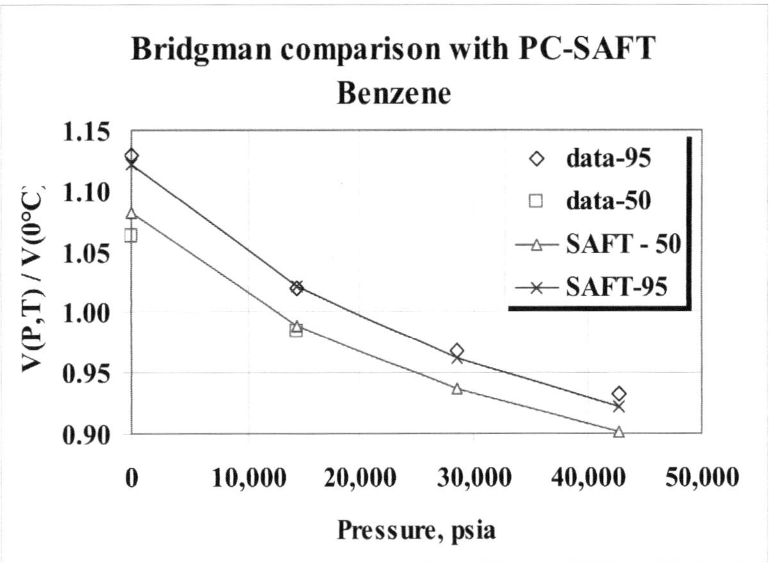

Figure 8: Comparison of PC-SAFT with P.W. Bridgman data for Benzene

While the heptane agreement is not as good as desired the benzene data agreement is good.

In conclusion the cubic equations of state are widely in simulations for petroleum exploration and production. These simple ready to used equations have been a nice balance between accuracy and simplicity. These equations will continue to serve the industry as new equations are developed for the frontier of developments where better pressure and temperature dependence are required.

Gross, J.; Sadowski, G., - Perturbed-Chaim SAFT: "An Equation of State Based on a Perturbation Theory for Chain Molecules," Ind. Eng. Chem. Res. 2001, 40, 1244-1260.

W.G. Chapman, K.E. Gubbins, G. Jackson, and M. Radosz, "SAFT: Equation- of-State Solution Model for Associating Fluids," Fluid Phase Equilibria, **52**, 31-38 (1989).

237f The Role of Experiments In the Development of Chemical Technology

Joan F. Brennecke, Department of Chemical and Biomolecular Engineering, University of Notre Dame, 180 Fitzpatrick Hall, Notre Dame, IN 46556

Tremendous strides have been made in the prediction of thermodynamic and transport properties from molecular simulations and theoretical models. Nonetheless, experimentation plays a critical role in both the conception of new chemical processes and their development and design. This is especially true when developing or exploring unconventional materials. We will give a number of examples of the importance of experimentation from our work with ionic liquids, which are organic salts with low melting points. We will show how serendipitous experiments led to the development of ionic liquids for gas separations and the use of carbon dioxide to separate mixtures of ionic liquids and organics. In addition, we will show how experimental results have been key to evaluating thermodynamic models of phase behavior involving ionic liquids.

History of the Clean Air Act

Prepared by Jason Templin, candidate for LL.M. degree, for
Lee Paddock, Associate Dean for Environmental Law Studies,
The George Washington University Law School

To understand the purposes of and methods behind regulation under the Clean Air Act, one must look at its origins in the common law of nuisance and its direct precursors in municipal ordinances. And any discussion of the common law must begin in England.

The first effort to control air pollution occurred in 14th century England. At that time, sea coal was used to fuel lime kilns. Combustion of sea coal produced a noxious brew of sulfur dioxide, carbon dioxide, and particulate matter emissions. Not coincidentally, as sea coal use increased, England's air quality became intolerable. To remedy this problem, King Edward I banned the use of sea coal in 1306, declaring that "all but smiths [shall] eschew the obnoxious material." He also urged his subjects to return to wood ("the fuel they used of old"). But the king's edict was ignored, and sea coal use continued unabated.

Air pollution continued to bedevil England in the ensuing centuries. As the country's population increased, so too did coal combustion. And with increased combustion, the air quality became even more degraded. Yet there were no air pollution laws to enforce. Instead, people had to sue responsible parties for injuries to themselves, their families, or their property. These "nuisance" claims enabled people to seek damages for their injuries and, in some cases, even an order enjoining the pollution.

The seminal nuisance case was *Aldred v. Benton* ("Aldred's Case"). Aldred sued his neighbor, a pig farmer, alleging that the "fetid and unwholesome stink" made his land unfit for occupation. Aldred was awarded damages, but Benton appealed, arguing that "it is lawful for anyone to make a hog-sty, even in a market town, for one cannot be so tendernosed." After losing his appeal, Benton went to the King's Bench. There, he argued that his right to earn a living trumped Aldred's aversion to noxious odors. But the King's Bench was not persuaded. It found that Benton had a duty to use his land in a manner that would not harm his neighbor.

Because Benton violated this duty, he was liable for damages. Thus, the court adopted an absolute liability approach to nuisance claims. It was an auspicious start for plaintiffs.

But the use of nuisance claims to control air pollution was short-lived. During the industrial revolution, nuisance claims became increasingly difficult to win. This was so for two reasons. First, plaintiffs had to prove that the defendant's noxious vapors caused their injuries and that the pollutant came from defendant's chimney. In the 1800's, there was only circumstantial evidence that noxious vapors caused illness (in fact, some doctors even opined that pollution was beneficial, since it might be used to fight cholera). And second, as industry spread, it became almost impossible to trace a pollutant to its source. Consequently, plaintiffs often faced insurmountable evidentiary obstacles in nuisance cases.

Plaintiffs also had to contend with less sympathetic courts. Unlike the court in *Aldred's Case*, which seemed to hold that a plaintiff's right to clean air was sacrosanct, 18th century courts were reluctant to award damages or injunctions to urban plaintiffs. Most courts believed that plaintiffs impliedly consented to noxious vapors when they lived in an industrial area. And if plaintiffs consented to this exposure, then they could not seek redress for their injuries.

Even more ominously for plaintiffs, courts in the 1850's began dismissing nuisance complaints as "trifling inconveniences." To overcome this bias, plaintiffs had to demonstrate that their alleged injuries were more than just irritations or annoyances. In most cases, plaintiffs could not show the requisite harm.

In America, nuisance claims arose in the heart of coal country, western Pennsylvania. From the beginning, these claims were difficult to prosecute. *Richard's Appeal* was illustrative of nuisance claims at this time. Richards, a cotton cloth manufacturer, sued an adjacent iron company that had built a smelting plant in the area. Richards claimed that smoke had "sulphurous, unwholesome, and noxious" effect on his family and property. He sought an injunction to order the company to build taller stacks or to use cleaner-burning coal. The trial court ruled for defendant, and Richards appealed. The Pennsylvania Supreme Court affirmed, finding that Richards failed to prove "feasible means of abatement" but suggested that Richards might be entitled to common law damages.

Four years later, in *Huckenstine's Appeal*, plaintiff sued a brickyard, alleging that noxious vapors from the brickyard harmed his orchards and that the conditions made the "land and the houses on it very inconvenient and uncomfortable as a residence." The trial court found for plaintiff and issued an injunction. But the Pennsylvania Supreme Court reversed. It held that plaintiff was not entitled to redress for injury to his "comfort." The court also found that plaintiff had consented to the vapors/odors of the area where he lived. It stated that "the people who live in such a city or within its sphere of influence do so of choice, and they voluntarily subject themselves to its peculiarities and its discomforts, for the greater benefit they think they derive from their residence or their business there." Thus, implied consent to pollution exposure would bar nuisance claims in America, too.

But under the right circumstances, courts were willing to award damages and even injunctions. In *Pennsylvania Lead Company*, plaintiff sued lead smelter for emitting "offensive, noxious and poisonous gases, fumes and vapors." The Pennsylvania Supreme Court upheld the injunction, stating that "all intelligent persons are aware that lead vapors are poisonous, and this the more so as they are often, as in the case in had, accompanied with arsenic. In this matter we need not chemists and experts to teach us, for common experience is sufficient." The court also noted the importance of the lead smelter's location, which was "in the midst of a rich suburban valley, occupied by farms and country residences."

In many cases, however, courts were unwilling to enforce injunctions. Instead, courts chose to award money damages. But this was hardly fair compensation. Without an order to abate emissions, companies would continue to fill the atmosphere will noxious fumes that damaged property and caused illness. And even when courts upheld injunctions, companies could satisfy the injunction by investigating the problem without making any changes. In *Sullivan v. Jones & Loughlin Steel Co.*, for example, the Pennsylvania Supreme Court refused to uphold an injunction regarding pollution abatement technologies. Under the injunction, Jones & Loughlin was required to "investigate" available technologies to deal with its dust emissions. Several years later, with no change in emissions, Sullivan sought enforcement of the injunction. The lower court held that Jones & Loughlin was in contempt, but the Pennsylvania Supreme Court reversed. It found that the injunction only required Jones & Loughlin to consider measures to abate pollution. Because it did this, the court found that it had complied with the injunction. But without regulation, dust emissions continued unabated.

Because nuisance claims failed to remedy the pollution dilemma, many cities in the late 19th century began implementing ordinances to combat smoke emissions. These ordinances were the first "statutory" attempts to solve the intractable emissions problem. St. Louis implemented its smoke abatement ordinance in 1867. It required companies to build chimneys that were at least "twenty feet" above surrounding buildings and habitations. In 1869, Pittsburgh adopted a similar ordinance, prohibiting trains from using dirty (i.e. bituminous) coal. One reporter described Pittsburgh as "just visible trough the mingled smoke and mist, and every object in it is black. Smoke, smoke, smoke—everywhere smoke!" These ordinances were designed to force companies to adopt new technology to reduce emissions. Although this approach would prove effective and be codified in the Clean Air Act and state legislation in the 20th century, it was largely ineffectual at the time. For without adequate enforcement, these ordinances failed to improve air quality and were largely symbolic.

Ten years later, Chicago passed an ordinance that made smoke "negligence per se." This obviated the need for plaintiffs to prove causation and trace the pollutant to its source. But there were still challenges. First, the ordinance only banned "dense smoke." Thus, plaintiffs had to prove that defendant's smoke was sufficiently thick. And second, the burden of enforcement was on a small group of inspectors. In many cities, smoke-emitting factories were too numerous to regulate with a few city workers. This meant that the ordinance was only as effective as its public support.

And public support was notoriously fickle. One writer described this period as one of "perpetual mobilization." Activists—mostly women—would galvanize the public and, for a time, convince people that anti-smoke ordinances were the answer. But over time, people lost interest. Moreover, some companies were able to invalidate these ordinances through the courts, since many courts considered such ordinances beyond the municipality's authority. In short, without adequate enforcement, municipal ordinances could never solve the nation's growing pollution problem.

During the early 20th century, sporadic pollution efforts continued. But a series of pollution disasters, both in the United States and abroad, changed public perception about emissions and were the impetus for federal legislation. In 1930, a deadly "fog" killed 60 people in Belgium. London had two smog incidents—one in 1948 and the second in 1952—that killed more than 4,000 people. And in 1950, "fog-bound fumes" from an oil refinery in Poza Rica, Mexico killed 22 people.

But for Americans, the most significant event occurred in 1948 in Donora, Pennsylvania. Donora was located on the Monongahela River in western Pennsylvania. In the 1940's, it was sandwiched between a zinc smelter and U.S. Steel's massive steel mills. During one week in October 1948, there was a temperature inversion, which kept the polluted air close to the surface.

The pollutant cloud was so dense that people could not see the sun at noon. By the time the cloud dissipated, 20 people were dead and thousands more had been hospitalized with serious respiratory illnesses.

Because of Donora and other smog-related incidents, the federal government intervened in 1955 and passed the Air Pollution Control Research and Technical Assistance Act. Although the act did not regulate air pollution, it directed the Department of Health, Education, and Welfare, with the help of the Surgeon General, to administer a funding program for state and local governments. This program kept pollution control at the state level but provided incentives for local governments to use new technologies to control pollution.

In 1963, Congress passed the Clean Air Act. The act gave Congress a greater role in air pollution control in two respects. First, the act authorized federal investigation if there was interstate air pollution. And second, the act authorized Congress to pass interstate compacts regarding air pollution. Still, pollution control was left largely to the states and local governments.

Four years later, Congress passed the Air Quality Act of 1967. The act focused on "stationary" sources of pollution and had four goals: 1) protect and enhance the quality of the nation's air resources so as to promote the public health and welfare and the productive capacity of its population; 2) initiate and accelerate a national research and development program to achieve prevention and control of air pollution; 3) provide technical and financial assistance to state and local governments in connection with the development and execution of their air pollution prevention and control programs; and 4) encourage and assist in the development and operation of regional air pollution control programs. To achieve these goals, Congress ordered the creation of "criteria" to protect public health. The act also provided for the eventual creation of ambient air quality standards.

In 1970, Congress amended the Clean Air Act in the Clean Air Act Amendments of 1970. These amendments are collectively known as "The Clean Air Act." To facilitate air quality improvements, Congress vested what had traditionally been state power in the Environmental Protection Agency (EPA). Now, EPA had the responsibility of establishing air quality standards for the nation. Each state, in turn, would implement these national standards through its State Implementation Plan (SIP).

Since 1970, the Clean Air Act has been amended twice. In 1977, Congress addressed "nonattainment" areas (i.e., areas that were still not in compliance, or attainment, with air quality standards). In addition to requiring states to designate "nonattainment" areas, Congress directed states to assess their SIPs regularly to ensure effectiveness.

In 1990, Congress amended the Clean Air Act to require states to designate areas as "attainment," "nonattainment," or "unclassifiable." The '90 amendments also set deadlines for state compliance with National Ambient Air Quality Standards (NAAQS) and committed EPA assistance on tropospheric ozone compliance.

Since the 14th century, people have been trying to deal with air pollution. Over the intervening centuries, states and countries have adopted many ways to regulate noxious vapors, some effective and others not. The Clean Air Act is an amalgamation of these approaches. It recognizes that air pollutants are a "nuisance" and, thus, must be regulated. But it also emphasizes fairness in encouraging companies to adopt the best available technology, which harkens back to municipal ordinances in the 19th and 20th centuries. Problems remain, to be sure. But by adopting the time-tested approaches to pollution regulation, Congress has equipped EPA and the states to deal with our air pollution in a fair and effective way.

254b The 100 Year War: A History of U.S. Air Pollution Science and Policy

John Bachmann, (Formerly Associate Director for Science/Policy at EPA's Office of Air Quality Planning and Standards), Vision Air Consulting, LLC, 120 Glen Ridge Dr, Chapel Hill, NC 27516

This presentation spans more than 100 years of U.S. air pollution history in an effort to illuminate how we arrived at the particular approaches to managing air quality reflected in the Clean Air Act (CAA). Today, we can take pride in the remarkable progress and benefits that have come from the efforts of the many participants in U.S. air quality management, particularly since the passage of the landmark 1970 CAA Amendments. Analyses have found that the estimated benefits of the CAA substantially exceed compliance costs, and the ratio of benefits to costs has actually increased with time. Yet, the CAA has been far from perfect and implementation has been subject to numerous delays and inefficiencies.

Since 1970, air quality management in the United States has depended in large measure on establishing effects-based National Ambient Air Quality Standards (NAAQS) that drive state and local programs to attain and maintain acceptable air quality. But what were the origins of air quality standards and management, and how did the United States come to adopt this approach for federal legislation in 1970? Why not other approaches, such as leaving air pollution control to the states or local governments, setting best technology limits as in Germany, or adopting economic incentives such as emission taxes? In addition to shedding some light on these questions, the presentation will focus on how NAAQS were established and reviewed after 1970. It shows how the NAAQS have motivated implementation programs, motor vehicle standards, regional emissions reductions, and how NAAQS-inspired research in effects, air quality, and control sciences influenced policy and legislation that improved the effectiveness of air quality management. The presentation concludes by looking toward the future of air quality management, emphasizing continuing and new challenges, particularly how to coordinate with climate change programs.

The miracles of science

DuPont SHE Excellence Center
1007 Market Street
Wilmington, DE 19898
September 22, 2008

AIChE's 2008 Annual Meeting taking place November 16-21, 2008 in Philadelphia, PA

Extended Abstract

DuPont has a long history of aggressively reducing our air emissions beyond compliance, whether it is greenhouse gases, which we reduced globally by 72% between 1990 and 2004 and continue to reduce, or air toxics which we reduced by 80%.

The Clean Air Act(CAA) has made significant contributions to improving America's environmental quality. Ambient air has improved continuously for the last 25 years and continues to improve.
Major improvements have been made in mobile sources and more reductions are coming as the fleet turns over. Electric utilities have made major reductions form the acid rain requirements and NOx SIP call. We are a manufacturing entity that operates small power facilities that are a much different scale than utilities, and our perspectives relate to those operations, not bulk power generation.

DuPont supported the Clean Air Act as approved by the 1990 amendments, we now have the opportunity to build on the experience since then to move toward improvement to eliminate barriers to improvement, streamline governmental processes and reduce the unnecessary ongoing costs to society while ensuring the high level of air quality the Act has helped to provide.

1. The National Ambient Standards (NAAQS) have generally been increased in stringency since 1990 and have made important contributions to air quality. Now though, we have reached a point where there is disagreement about the basic interpretation and methods of data and science evaluations being conducted and on their significance levels. Some levels now approach natural background levels. The 5 year renewal time is much too short to allow the EPA and other researchers to adequately do sound scientific reviews with input from all stakeholders. A 10 year NAAQS review period would be more reasonable at this point in time.

CAIR/CAM: The vacature of CAIR and CAMR creates a pretty big gap relative to NAAQS compliance and regional haze. There needs to be some way to address further utility unit emissions reductions in an economically practical manner.

2. The Operating permit program in Title V has been much more expensive than anticipated and fees continue to escalate. We are concerned that the fees are now out of proportion and result in a significant cost burden on entities at a time when high energy and health care costs challenge the global competitiveness of US manufacturers. For example, at DuPont we have sites where Title V fees exceed $500 thousand per year. A reduction is possible if the renewal period was modifies from the current 5 year period to a longer period or if permit modifications were triggered only by substantive and significant changes too an operation.

The current system should also be streamlined by having only one permit modification procedure, public hearings etc for NSR/PSD vs. the current duplicative system.

3. The current NSR/PSD program can represent significant time delay and expensive procedure for sources to obtain permits, including for projects that improve energy efficiency and reduce greenhouse gas emissions and these extended timeframes can further inhibit competitiveness in a world where businesses have to be nimble.

The program could be improved through implementation for existing sources of programs like acid rain and the Air toxics residual risk programs where need is determined by local/regional conditions and air quality evaluation. The modification rules as implemented by EPA in the past have become a constant litigation and interpretative nightmare. The 2002 rulemakings for the arithmetic test for determining emission increases from modifications are a significant improvement over the emission increase methods previously used by EPA enforcement.

4. The trend toward expensive hardware to continuously monitor emissions with all the continuous evaluations required for small sources such as typical in the manufacturing world (i.e. less than electric utility size) to allow a variety of different techniques to be used for determining emissions. Sources typically spend over $100,000 annually per regulated monitor for NOx or SOx continuous monitors for large size industrial boilers. The cost is the same regardless of the size of the emissions. The cost per ton of emissions skyrockets as each emission point becomes smaller. Industrial cost of monitoring can easily exceed electric utility cost of control including continuous monitoring.

These smaller sources need to be able to use available information in most cases, with invoices as backup support for fuel purchases. Fuel metering as for natural gas is usually pretty good. Fuel monitoring for coal is not possible unless you have gravimetric feeders (none of our units have them), so we use calculated fuel use, coal purchases invoices, inventory corrections, etc. combined with fuel analysis and emission factors.

5. DuPont has long been reducing our own greenhouse gas emissions, and we are actively advocating for comprehensive, mandatory US legislation to reduce greenhouse gas emissions. We think that to deliver the needed reductions in an economically sustainable manner that greenhouse gas reduction program needs to be intentionally developed, and that simply adapting the elements of the current Clean Air Act to greenhouse gas controls without Congress adapting them would be ineffective and costly.

SUMMARY:
DuPont supports a robust Clean Air Act as well as reductions in greenhouse gas emissions. We believe that greenhouse gases should be addressed through a new construct developed by the US Congress for that purpose. We also believe we should look for opportunities to increase the efficiency of the existing CAA while ensuring its ongoing effectiveness.

E. I. duPont de Nemours and Company

254d Preparing for Potential Greenhouse Gas Legislation

Sunil P. Hangal, Environmental Compliance Assurance Management (ECAM), 9 Smoke Rise Lane, Bedminster, NJ 07921

Based on the commitment of both the presidential candidates and the regional initiatives across the US, it appears that a national Carbon Cap and Trade System may be legislated within the next presidential term. This presentation will focus on the following:

1. California AB 32 rulemaking and California Air Resource Board (CARB) proposed plan for Cap and Trade System.

2. The various state initiatives including the Regional Greenhouse Gas Initiative (RGGI) by North Eastern and Mid Atlantic States.

3. A brief introduction to the Kyoto Protocol, European Union Emission Trading System (EU ETS) and the United Nations Clean Development Mechanism (CDM).

4. The voluntary carbon market currently operating in the US. Offset creation and use for voluntary and compliance purpose.

5. The key issues that law makers in the US are likely to address in any future legislation.

6. Key points of a Cap and Trade System Design – Allocations, linkages, price caps and offsets.

7. Early action companies can take to get benefits from any potential Carbon legislation and address concerns from investors and customers.

8. How some US companies are taking actions to prepare for Carbon legislation and market.

259a 40 Years of Surface Science at Stanford: A Kinetic Perspective

Robert J. Madix, School of Engineering and Applied Sciences, Harvard University, Pierce Hall, Cambridge, MA 02138

Appreciable advances in understanding catalytic cycles have been advanced by surface science studies of reactivity. In this talk a mechanistic and kinetic perspective will be offered relating work spanning four decades to principles related to catalytic reactions on metals.

259b Toward Ultimate Selectivity: Site-Directed Chemistry and Catalysis at Pt-Sn Alloy Surfaces

Bruce Koel, Chemistry, Lehigh University, 6 E. Packer Ave., Bethlehem, PA 18015

Bimetallic catalysts involving Sn and Pt have important applications in hydrocarbon conversion catalysis, but little fundamental information is known about the chemistry and catalysis that occurs at distinct Pt-Sn alloy phases, how this chemistry varies with composition and structure, whether various phases are optimal or detrimental to catalyst performance, and whether novel nanostructures have special chemical properties. The need for improved selectivity of catalysts, and improvements in catalyst performance overall, drive our studies. We have carried out a wide range of investigations probing chemisorption and reaction kinetics on well-defined, bimetallic Pt-Sn surfaces in order to improve basic understanding of the chemistry of hydrocarbon molecules and reactive intermediates on such surfaces. I will summarize our results, focusing on selective dehydrogenation and hydrogenation reactions, and briefly discuss more generally the role of a second, added metal component in chemistry and catalysis on alloy surfaces.

259c Surface Science Studies of Metal Oxides

John M. Vohs, Chemical and Biomolecular Engineering, University of Pennsylvania, Room 311A Towne Building, 220 South and 33rd Street, Philadelphia, PA 19104

Ultra-high vacuum spectroscopic techniques have proven to be valuable tools in elucidating the relationships between the structure and reactivity of surfaces. At least for catalytic systems, work in this area has been dominated by studies of the surface properties of metals. Over the last twenty years, however, surface science studies of catalytically active metal oxides have emerged and are now almost as prevalent as those of metals. In this talk I will give a historical perspective of surface science studies of metal oxides as well as describe the work of my own group in this area. Future challenges for studying the surface properties of oxides will also be presented.

259d Gold Atom Release In Surface Oxidation Reactions

Cynthia Friend[1], Xiaoying Liu[1], Baker Thomas[1], and Weiwei Gao[2]. (1) Chemistry and Chemical Biology, Harvard University, Cambridge, MA 02138, (2) Department of Chemistry and Chemical Biology, Harvard University, 12 Oxford Street, Cambridge, MA 02138

Oxidation of alcohols and olefins on extended Au surfaces will be discussed and related to high-surface area catalysis. A clear connection between mechanisms in ultrahigh vacuum to those under operating conditions will be illustrated. Density functional theory calculations that provide insight into the bonding and reactivity of various defective structures on the surface will also be discussed.

291a Historical Perspectives on Chemistry and Chemical Process Development and Innovation at Dow

Victor Atiemo-Obeng, Rik Tuinstra, and Jack Kruper, Jr. Research & Engineering Sciences, The Dow Chemical Company, 1776 Bldg., Midland, MI 48674

The Dow Chemical Company is a leader among a handful of global integrated chemical companies. The building of this leading chemical company from its early beginnings in 1897 embodies a rich history of chemical process development and innovations. In this presentation a simplified structure of the interrelated chemistries and chemical processes that are practiced will be outlined. The various drivers for innovation that influence the selection of specific chemistries and processes will be discussed.

291b Ernest W. Thiele: His Impact on Chemical Engineering

Jonathan Worstell, Hoder, Shell Global Solutions US, 3333 Highway 6 South, Room CP-109, Houston, TX 77082

Ernest W. Thiele impacted and provided guidance to three areas of chemical engineering during each of their infancies. Those areas were:

distillation;

solid porous catalysis;

nuclear reactions and nuclear engineering.

This talk presents an appreciation of Thiele's career and discusses the impact Thiele had on distillation, catalysis, and nuclear engineering --- before it became nuclear engineering, when it was still chemical engineering.

291c DuPont: Building on the Past towards a Sustainable Future

Ray W. Miller, DuPont Applied BioSciences, 4417 Lancaster Pike, Building 728/room 1419, Wilmington, DE 19805

DuPont, one of the world's oldest leading chemical giants, is rapidly becoming an industrial biotechnology powerhouse. Ray will explain the driving forces behind this transformation, and highlight DuPont's development and commercialization of sustainable materials and energy offerings starting over a decade ago, before higher oil prices and global warming trends were widely recognized concerns. He will focus heavily on the development and commercial introduction of the Bio-PDOÔ and Sorona" polymer platforms and progress being made to broaden the offerings with value-added specialties such as Cerenol™ based on this initial platform molecule (bio-sourced 1,3 propanediol). The presentation will also briefly touch on the DuPont DOE-matching funds bio-refinery program and DuPont's partnership with BP to introduce biobutanolÔ fuel for the future as well as other applications of biosciences in the surfaces and medical areas. The presentation will conclude with some suggestions on the role of past business models in helping achieve these fundamental shifts away from petroleum dependency in our materials and fuels sectors.

Web Page: www.renewable.dupont.com

Shell Hydroformylation (SHF)

Joseph B. Powell, Ph.D.
Chief Scientist – Chemical Engineering
Shell Global Solutions U.S. Inc.
Westhollow Technology Center, Houston, TX

Hydroformylation or the "Oxo" reaction was identified by Otto Roelen in 1938 as an explanation for oxygenated byproducts during heterogeneously-catalyzed Fischer-Tropsch synthesis.[1,2] The observed addition of carbon monoxide and hydrogen to olefins to form aldehyde and alcohol products was found to occur in the liquid phase, catalyzed by soluble cobalt carbonyl complexes leached from the heterogeneous cobalt catalyst under reaction conditions. The discovery ushered in a new era of "organometallic homogeneous catalysis". In the early 1960's, Slaugh and Mullineaux[3] of Shell Oil discovered that addition of phosphines (σ-donors) provided for improved catalyst stability at low pressures. Ligand modification also provided the abilities to hydrogenate aldehyde intermediates to alcohols in a single step, and to control reaction selectivity via steric impacts.[4] The phosphine-modified cobalt-catalyzed process became known as "Shell Hydroformylation" or SHF. The concept of ligand-modification was subsequently extended to catalysis by rhodium carbonyls for low-pressure hydroformylation.[5-7]

The ground-breaking Shell Higher Olefins Process (SHOP) made its debut in the late 1970s in the US. It opened up a new route to olefinic raw materials which serve as feedstock for SHF processes, to support the growing detergent industry. Detergent alcohol SHF products are used by customers to manufacture personal care products, soaps, shampoos and household cleaning solutions. Current Shell production is estimated at 1.2 billion pounds per year.

In 1999, SHF technology was leveraged to the hydroformylation of ethylene oxide[8] to produce 1,3-propanediol (PDO), a monomer for production of a novel polyester polytrimethylene terephthalate (PTT). PDO nameplate production capacity is 160 million pounds a year. The new PTT polymer is finding commercial application in carpets and apparel.

References

1. J. Falbe, *Carbon Monoxide in Organic Synthesis*, Springer-Verlag, NY (1970).
2. R. F. Heck and D. S. Breslow, *J. Am. Chem. Soc., 83*, 4023 (1961).
3. L. H. Slaugh and R. D. Mullineaux, *J. Organometallic Chemistry*, *13,* 469 (1968); U.S. Patents 3,239,569 and 3,239,570 (1966).
4. C. A. Tolman, "Steric Effects of Phosphorus Ligands in Organometallic Chemistry and Homogeneous Catalysis", *Chem. Rev. 77,* 313 (1977).
5. L. H. Slaugh and R. D. Mullineaux, US Patent 3,239,566 (1966).
6. D. Evans, J. A. Osborne, and G. Wilkinson, "Hydroformylation of alkenes by use of rhodium complex catalysts", *J. Chem. Soc. A. 33 (21): 3133–3142 1968).*
7. R. L Pruett and J. A. Smith, *J. Org. Chem.*, *34*, 327 (1969).
8. J. B. Powell, L. H. Slaugh, T. C. Forschner, T. B. Thomason, T. C. Semple, P. R. Weider, J. P. Arhancet, US Patents 5,463,144; 5,463,145; 5,463,146 (1995).

291e Past, Present, and Future of Gas to Liquids Technology

Sagar B. Gadewar[1], Daniel J. Auerbach[1], Eric W. McFarland[2], and Vivek Julka[1]. (1) GRT Inc., 861 Ward Drive, Santa Barbara, CA 93111, (2) Chemical Engineering, University of California, Santa Barbara, Engineering 2, Room 3357, Santa Barbara, CA 93106

The development of technology for the economical conversion of natural gas to easily transportable liquid products (Gas-to-Liquids, GTL technology) has a rich history and yet remains a challenging and important problem for Chemical Engineering. GTL technology was first developed in the 1920s and made use of the Fischer-Tropsch (FT) process to convert synthesis gas (Syngas) to mix of linear alkanes suitable for diesel fuel. Sasol in South Africa has been at the forefront in the application of the Fischer Tropsch technology, using coal and natural gas as a feedstock for Syngas formation. Modifications to the basic Syngas-FT process have been made by Exxon-Mobil, Shell, and others who have produced low-sulfur diesel fuel. Methanol is also readily synthesized as a liquid product from natural gas derived Syngas and Exxon-Mobil's process to convert natural gas into gasoline made use of the methanol intermediate. At present, methanol is the most widely produced liquid product from natural gas, however, several large commercial projects have been recently completed or are under construction for Syngas-FT based GTL. In recent years, increasing oil prices and the availability of large quantities of relatively low cost and underutilized natural gas have driven an increased interest in improving GTL technology, particularly to deal with natural gas resources that cannot currently be developed economically. New process development methodologies and tools play an important role in developing the new generation of GTL processes. We will discuss several new approaches to this problem. Technology developed by GRT Inc. using bromine activation of natural gas allows small scale production of liquid hydrocarbons such as gasoline and aromatics from stranded gas reserves. This new approach can address the problem of monetizing small gas resources and can reduce or eliminate flaring in small to medium size gas fields.

Web Page: www.grt-inc.com

291f The Challenges of Diminishing Natural Capital in the Transition to a Bio-Based Industry

Luca Zullo, Environmental Finance, Cargill Incorporated, 12700 Whitewater Drive, Minnetonka, MN 55343 and Deborah Ross, Process Solutions, Cargill Incorporated, 12700 Whitewater Drive, Minnetonka, MN 55343.

For most of human history, global economic activity has been largely derived from the exploitation of the earth's biosphere via agricultural processes. It is not incidental that the motto of the US Department of Agriculture, which was established during the Lincoln administration, is "The source of all industry and commerce". However, due to the abundance of natural capital, the earth's resources have historically not been considered or translated into economic terms as real capital.

Today, the fact that most of the world's industrial economy is dependent on finite fossil resources causes great concern due to supply as well as the numerous negative environmental impacts of fossil fuels. This concern has catalyzed an irreversible trend toward exploiting the biosphere with more activities than simple food production. Developed economies are now demanding biofuels and bio-based materials from agricultural systems while developing nations are increasing their demand for food. In addition, agriculture itself has become enmeshed with the "unsustainable" industrial systems of the XX century due to its dependency on fossil fuels in transport and farming and the use of synthetic fertilizers mainly derived from natural gas. In concert, these trends are causing unprecedented strain on the earth's biosphere as well as driving commodity prices to record highs. Biofuels, which were once heralded as the answer to energy and climate change issues, are now being blamed for rising commodity prices and agricultural strains.

The limits of "natural capital" are evident and there is now a new appreciation for the limits of the biosphere. It is our argument that there is a bright future for the bio-based industry if we accept the real cost of natural capital and we nurture it accordingly. In this talk, we will review present and future prospective for the bio-processing industry; with an emphasis on bio-fuels and biomass. We will argue that while the trend is not reversible, a systems approach of process engineering needs to be applied to the biosphere of production at a macro-scale and not only inside the plant gate to deliver the promise that these technologies hold. This approach enables unique opportunities for new avenues of academic research and industrial applications to responsibly leverage natural capital in a global economy.

306a Water Quality Regulation - the Past 100 Years

Steven T. Miano, Hangley Aronchick Segal & Pudlin, One Logan Square, 18th & Cherry Streets, 27th Floor, Philadelphia, PA 19103-6933

The scope and breadth of water quality regulation has changed significantly over the last 100 years. These changes have, in most cases, been reactive. Industrial activities at the turn of the century turned rivers into industrial sewers. Rampant disposal of industrial wastes into rivers continued through the mid 1900s, culminating in rivers such as the Cuyahoga in Cleveland, OH catching fire in 1969. These activities led, inevitably, to more stringent water quality regulation. In addition, future population shifts, agricultural activities and global warming will continue to have significant effects on water quality. This session will discuss the history of water quality regulation, tracing its roots from the 1899 Rivers and Harbors Act, to the Clean Water Act of 1948, the Federal Water Pollution Control Act Amendments passed in the 1970s, the Safe Drinking Water Act in 1974, the Oil Pollution Act of 1990, and subsequent amendments to most of these statutes. In addition, many states had independent, and sometimes earlier, regulatory programs that played and still play a significant role. Also, law suits by environmental groups forced further regulation required by statute but never implemented by regulatory agencies. Examples include wetlands regulation, stormwater and Total Maximum Daily Loads (TMDL). Finally, recent litigation before the US Supreme Court has resulted in significant uncertainty regarding federal jurisdiction of waters of the United States. The historical perspectives gained during this session will provide a solid foundation for understanding and working with likely future water quality regulation.

306b Water Quantity Regulation - the Past 100 Years

Joseph W. Dellapenna, School of Law, University of Villanova, 299 North Spring Mill Road, Villanova, PA 19085-1682

Allocation of water resources has been and continues to be a very significant topic in the United States. Although historically considered a concern primarily for western states, eastern water supplies have become increasingly erratic due to overuse, short-term droughts, and potential long-term climatic changes. These factors have been aggravated by development patterns, groundwater pollution and saltwater intrusion. Interestingly, epic western water wars have recently come east and there is no end in sight. This session will discuss the history of water quantity regulation in both the western and eastern portions of the United States, from English common law through increasing statutory regulation. It will also provide insight into the directions both eastern and western states have taken and likely will continue to take in confronting ever-increasing shortages of water.

306c Interstate Water Regulation - Historical, Current and Future Direction

Kenneth J. Warren, Hangley, Aronchick, Segal & Pudlin, One Logan Square, 18th & Cherry Streets, 27th Floor, Philadelphia, PA 19103-6933

The regulation of water use has and continues to be the province of the states. However, the flow of water, whether in surface water bodies or under ground, does not respect political boundaries. Water flows freely across state and local borders. As a result, certain states have banded together, in some cases with the federal government, and formed interstate compacts to regulate water use. Examples include the 1961 Delaware River Basin Commission in Pennsylvania and the 1922 Colorado River Compact in the west. There are many others throughout the United States. While each compact has its own history and purpose, all share the common goal of attempting to meaningfully regulate water use across state boundaries. Focusing on the Delaware River Basin Commission, this session will provide an overview of the jurisdiction, activities and accomplishments of one of the most successful interstate water compacts and will provide useful insights into the critical role played by interstate regulation of water use.

The Role of Science and Engineering in Water Regulation Over the Past 100 Years

James R. May*
Patrick Clary#

Abstract

Scientific and engineering principles are inexorably linked to the regulation of water. Scientists and engineers first discovered the link between disease and water sources in the mid-19th century. Over the years, scientists and engineers have led the way to identifying water quality problems and their causes. These discoveries have directly contributed to the scope of water regulation in the United States and elsewhere. In addition, changes in water quality regulation have dictated the need for increasingly sophisticated water treatment technologies and engineers have been at the forefront of the development of these water control technologies. This session will discuss the historical roots, and the development over the century, of scientific principles and engineering technology in identifying and combating water pollution. It will also provide a solid foundation for understanding current efforts underway to solve current and future water quality issues.

Introduction

Scientists and Engineers have played a central role in shaping water regulation over the last 100 years. Prior to about 100 ago, human waste was commonly discarded into open ditches, sinks and gutters. This, of course, diminished the quality of life, particularly in urban centers, and helped to spread disease.

Thus, cities (and their planners and engineers) began to design sewer and other sanitary systems to transport human waste for discharge into nearby water bodies, often used for drinking. This led to further spread of disease, including typhoid due to discharge of raw sewage into these sources of drinking water.

Scientific advancements and the discovery of germ theory changed this practice. This led in turn to discharge of raw sewage into surface waters, including ocean waters, estuaries, lakes, rivers and streams not directly used for drinking. This once again led to concerns about the spread of disease from urban centers elsewhere, including those from scientists that human waste from New York City caused or contributed to the spread of disease in New Jersey, or that Chicago's wastes caused sickness and death in St. Louis and elsewhere.[1]

---

* James R. May is a Professor of Law and Adjunct Professor of Engineering at Widener University

Patrick Clary is a student at Widener University School of Law in Wilmington, Delaware

[1] New York v. New Jersey, 256 U.S. 296 (1921); Missouri v. Illinois, 200 U.S. 496 (1906).

Over the last 100 years, scientists and engineers have played a central role in shaping water quality goals. Engineers are largely employed to design technological systems to control pollution so as achieve water quality goals. Scientists are largely enlisted to inform discussions about the level of pollutant stressors that receiving waters can withstand and still be sufficiently safe for aquatic and human use.

During the evolution of America's regulation of water pollution, law, from the first permit systems at the beginning of the twentieth century, to the modern day Clean Water Act, the sciences and engineering have played an integral role. The large-scale management of water requires large-scale infrastructure. Nowhere have the sciences and engineering been more instrumental in protecting water than in the area of water treatment and delivery systems. However, the sciences and engineering have done more than simply design our piping and filter our drinking water. Scientists and engineers are an integral element in the regulatory process. When disagreements over water regulation spill into the courts, scientists and engineers also provide expert testimony and help settle disputes. Water regulation would be impotent without the sciences and engineering.

This paper discusses the evolution of engineering and the sciences in regulating water pollution during the last 100 years. First, it first traces the role of science and engineering in shaping water quality laws over the last century, culminating with the Clean Water Act, the nation's premier water pollution control law. Second, it examines the pivotal role engineers have played in developing the Clean Water Act's principal means of achieving water quality goals: technology based standards. Third, it discusses the role scientists play in developing and achieving water quality standards under the Clean Water Act. Fourth, it briefly accounts for how technology based standards and water quality standards are implemented under the Clean Water Act's permitting system. Fifth, it touches on the emerging role engineers and scientist play in addressing water rights – as opposed to water pollution – issues. Last, it describes the role of science and engineering in addressing ongoing water quality challenges. Essentially, law, science and engineering work in tandem to protect the strong public interest in clean water, and to track and control pollution so as to keep waters sufficiently safe and healthy for intended purposes.

Three's Company: Science, Engineering, Congress and the Clean Water Act

Clean, safe, fresh water is important to the nation's environment, economy, and security. If all of the water in the world were represented as 100 gallons, 97 would be undrinkable saltwater. Two gallons would be trapped in glaciers and icecaps. Two quarts would be groundwater. Less than one-half pint (the carton size of elementary school milk) would be freshwater. Of that one-half pint, roughly one teaspoon exists within a two hour drive from the locus of this conference, which consists of almost 60 million people, and draws an additional 50 million visitors who make 250 million trips annually, and spend $20 billion on services, $5 billion on tourism, $75 billion on food and fiber, and enjoy 3 billion pounds of fish and shellfish. Manufacturers and cattle and crop farmers need clean water too. The ones in this region alone use more than 15 billion gallons of water a year, producing products worth more than $25 billion.

The nation's great commercial and social traditions depend on ready access to clean, dependable water. A century ago, DuPont depended on the fast, clean-flowing Brandywine River to fuel his first powder mills and usher in the petrochemical revolution. Today, clean water is no less important than it was then. It goes without saying that clean water is important for fish and wildlife and their habitat. What we often forget is that clean water is essential to the nation's great petrochemical, pharmaceutical, agricultural and animal farming, mining, automotive, port, fishing, crabbing, housing, shopping, motel, restaurant and tourism industries. The nation's manufacturing sector depends on clean water to run efficiently. By maintaining property values, clean water is good for homeowners and developers. It helps keep health care, taxes and insurance costs down. In short, clean water is essential to the success of the nation's economy, sustains our values and quality of life, and is good for business. Following the Great Depression and World War Two, Congress sought to apply broach federal regulatory programs to address national challenges, from commerce to clean water. During this time, federal approaches to water pollution evolved away from focusing on local water conditions to one establishing national, technology based standards based upon feasible an engineering practices.

It took a while to submit to a technology based approach. Sixty years ago Congress enacted the Water Pollution Control Act of 1948[2], which established a kind of quasi state and federal cooperative to provide means for resolving disputes about interstate water quality. The Federal Water Pollution Control Act of 1956[3] continued this approach. Neither law did much if anything to improve water quality.

With the Water Quality Act of 1965[4] Congress had states set water quality standards, for example, maintaining 5 parts per million of oxygen in interstate freshwater used for trout fishing. The federal role was minimal, however, largely consigned to helping to resolve interstate disputes. Again, the 1965 Act resulted in little activity, and water quality continued to decline dramatically.

Indeed, at this time the majority of people in the country lived near water too polluted to use. One could not fish, swim or even boat in large parts of the Delaware and Schuylkill Rivers, suppressing commerce, home values and health. The Anacostia River was dying. Baltimore Harbor and the Port of Wilmington closed to all but tanker traffic, and major rivers like the Delaware and the Susquehenna were open sewers, dashing hopes of urban revitalization.

Congress considered how to improve water throughout the late 1960's and early 1970's. At core were two schools of thought about how to make the nation's waters "fishable and swimmable." The first, largely based in the Senate, would dispatch with water quality standards-based approaches in favor of a command and control, technology based approach led by the newly established EPA. The second, largely from the House of Representatives, would use a more muscular ambient based approach than contained in the 1965 Act. It had EPA set water quality criteria for certain "designated" uses, such as swimming, fishing, drinking or recreation. It then had the states – as had the 1965 Act – develop and implement water quality standards to meet designated uses. The compromise legislation largely embraced the Senate's technology-based approach, with consideration of local water quality conditions in

---

2 *See* 33 U.S.C. § 1251 – 1376 (2008).
3 *Id.*
4 *Id.*

those instances where installation of innovative technologies alone was not sufficient to achieve water quality standards.

The Clean Water Act of 1972[5] reflects a hybrid of the competing technology and ambient-based approaches. It is largely a response to earlier failed approaches rooted in common law that linked compliance responses to the ability of the water body to withstand the polluting activity. Congress determined that there is not sufficient scientific certainty to measure the "tolerable effects" of discharges.[6] Thus Congress opted for an innovative approach based on national, uniform technology based standards for categories and classes of point sources.

The Clean Water Act requires EPA to set technology-based standards for categories and classes of point source dischargers, such as pulp and paper mills, breweries and steam-electric fossil-fuel burning steam electric generating stations (power plants). When calculating these standards, regulators "aim to set effluent limits at feasible levels while still encouraging innovation."[7] It also allows EPA and states to set more stringent requirements wherever technology based standards alone are not sufficient to protect water quality for a particular water body. Engineers and scientists are involved every step of the way.

The Role of Engineers and Scientists in Shaping Regulation of Water Regulation

A. Role of Engineers in Shaping Water Quality

More than anything else, engineers help design, apply and implement technological means of controlling pollutant discharges as a means of protecting water quality. The following sections discuss some of the legal concepts that engineers and scientists apply in water regulation. For a listing of some selected scientific concepts and engineering responses to achieve water quality and protect drinking water systems, please see the appendix that appears at the end of this paper.

1. **The Use of Technology Based Standards**

Most federal environmental laws contain provisions that require EPA to set national technology-based standards on the regulated community, rejecting quasi-common law regulatory approaches that link the regulatory response to local conditions. For example, the first such statute to do so in 1970 – The Clean Air Act – requires EPA to develop and implement technology based standards for "stationary" and "mobile" of air pollution.[8] Since 1972, the Clean Water Act has required EPA to set and implement technology based standards for "point sources" of water pollution.[9] The Safe Drinking Water Act requires EPA to set technology based standards drinking water from "public drinking water systems."[10] Safe

---

[5] 33 U.S.C. § 1251 – 1376.

[6] EPA v. Cal. State Water Res. Control Bd., 426 U.S. 200, 202 (1976).

[7] *Id.* at 38.

[8] Clean Air Act of 1970, Pub. L. No. 91-604, 84 Stat. 1676 (codified as amended at 42 U.S.C. § 7401-7671g (1994)).

[9] Clean Water Act of 1972, Pub. L. No. 92-500, 86 Stat. 816 (codified as amended at 33 U.S.C. §§ 1251-1385 (1994)).

[10] Safe Drinking Water Act, 33 U.S.C. § 300g-1(b)(4) (2008).

Drinking Water Act, 33 U.S.C. § 300g-1(b)(4). The Resource Conservation and Recovery Act requires EPA to set technology based standards for facilities that dispose hazardous wastes.[11]

Congress has turned to technology-based standards because they are fair, predictable, efficient, adaptable, and enforceable. First, they are even-handed. Rather than allowing inefficiently disparate, site-specific standards, national technology based standards level the playing field by treating those who engage in regulated activity to meet a certain threshold standard. Second, they are predictable. Rather than being subjected to agency decision-making on an ad hoc basis, national technology based standards allow the regulated community to forecast regulatory requirements. Third, they are socially efficient. Regulatory rulemaking – even that taking many years as in the case at hand – is still generally a more efficient regulatory mechanism than case-by-case determinations. Fourth, they are adaptable. Technology based standards can be adjusted to fit innovation and feedback. Last, they are readily enforceable. Technology based standards are usually embodied by a particular technological requirement or a performance standard, either which are readily discernible by governmental enforcement agencies and by courts.[12]

2) **Water Pollution Control is Primarily Technology (Engineering) Based**

Technology-based standards are the central regulatory tool adopted in the Clean Water Act for achieving the "national goal that wherever attainable, an interim goal of water quality which provides for the protection and propagation of fish, shellfish, and wildlife and ... for recreation in and on the water."[13] Under this approach, Congress instructed industry to reduce environmental harms as much as technology allows. In so doing, Congress made a sharp break with the failed regulatory approach that had governed until 1972, which sought to meet water quality standards by trying to determine how much pollution the water could assimilate and still be clean enough for human and aquatic uses. Congress found this pre-1972 approach to be "inadequate in every vital respect."[14] Instead, technology based standards "facilitate enforcement by making it unnecessary to work backward from an overpolluted body of water to determine which point sources are responsible and which must be abated."[15]

Technology based standards apply to existing and new point sources. The level of performance for existing facilities is a function of pollutant, and becomes more stringent over time. Initially, existing point sources were to meet "best practicable control technology currently available" (BPT) for all regulated pollutants.[16] Thereafter, they were to meet more stringent technology based standards, as a function of whether the regulated pollutant is "conventional"

---

[11] Resource Conservation and Recovery Act, Pub. L. No. 94-580, 90 Stat. 2795 (codified as amended at 42 U.S.C. §§ 6901-6992K); *see* e.g., Patricia McCubbin, *The Risk in Technology Based Standards*, 16 Duke Envtl. L. & Pol'y F. 1, 7-8 (2005).

[12] Wendy Wagner, *The Triumph of Technology Based Standards*, 2000 U. Ill. L. Rev. 83, 87-105 (2000).

[13] 33 U.S.C. § 1311 (2008).

[14] *See* Milwaukee v. Illinois, 451 U.S. 304, 310 (1980), quoting S. Rep. No. 92-414, 7 (1971), 2 Legislative History of the Water Pollution Control Act Amendments of 1972, 1425 (Committee Print compiled for the Senate Committee on Public Works by the Library of Congress), Ser. No. 93-1 (1973).

[15] EPA v. Cal. ex rel. State Water Res. Control Bd., 126 U.S. 200, 204 (1976).

[16] 33 U.S.C. § 301(b)(1)(A) (2008).

(such as oil & grease, pH, nitrogen, phosphorus and dissolved oxygen) or non-conventional (such as toxics and metals). Conventional pollutants discharges were to comply with "best conventional pollutant control technology," (BCT).[17] Non-conventional pollutant discharges were to comply with "best available control technology economically achievable ... which will result in reasonable further progress toward the national goal of eliminating the discharge of all pollutants," (BAT).[18] New sources were to meet "best available demonstrated control technology" (BADT). EPA, or a state with delegated permitting authority, would then implement the applicable technology based standards when issuing National Pollutant Discharge Elimination System (NPDES) permits. The CWA prohibits the discharge of a "pollutant" from a point source in the absence of a permit.[19] EPA must set technology based standards for point source discharges of pollutants[20], including for discharges of "heat."[21]

3.) **Examples of Technology Based Standards Under the Clean Water Act**

a.) Best Practicable Technology

The Clean Water Act has a broad range of technology-based standards. Best Practicable Technology (BPT) is the least restrictive technology-based standard. Instead of setting forth a standard not widely available in an industry, BPT surveys existing technologies and selects those operating at the most efficient levels.[22] Facilities may adopt the specific technology used by the top-performing facilities, or develop an alternative that performs at the same level.[23] Compliance may therefore take the form of further innovation. As companies seek to lower the costs of modifying their facilities to meet the BPT standard, engineers and other scientific professionals assume a very important role.

With water regulation, these technology-based standards describe the desirable characteristics of water when it is released from some municipal or commercial facility. This type of regulation is called "effluent limitations," because the concern is the integrity of the water as it enters the nation's streams, lakes, and rivers. A selected few "effluent limitations" will be discussed to highlight how regulators create and enforce technology-based standards.

Water regulation touches many subjects, and in 1976 the EPA codified effluent limitations on the photography process.[24] This regulation is a Best Practicable Technology standard. These regulations describe a permissible daily discharge of certain contaminants (silver and carbon nitrate, as well as parameters for pH levels) that are used in the photography process. The regulation places an additional limitation on the average daily discharge for a 30 day period. These limitations are:[25]

---

[17] *Id.* § 301(b)(2)(E).

[18] *Id.* § 301(b)(2)(A)

[19] 33 U.S.C. § 1311 (2008).

[20] 33 U.S.C. § 1311 (existing sources) & 1316 (new sources).

[21] 33 U.S.C. § 1362(6) (2008).

[22] JOEL M. GROSS & LYNN DODGE, CLEAN WATER ACT 39-40 (American Bar Association 2005).

[23] LYNN M. GALLAGHER, CLEAN WATER HANDBOOK 44-45 (Government Institutes 2003).44-45.

[24] 40 C.F.R § 459 (2008).

[25] *Id.* § 459.12(a).

Table 1 BPT – 40 C.F.R § 459.12(a): Effluent Limitations on the Photography Process

Effluent characteristic	Effluent limitations	
	Maximum for any 1 day	**Average of daily values for 30 consecutive days shall not exceed—**
	Metric units (kilograms per 1,000 $m^2$of product)	
Ag	0.14	0.07
CN	0.18	0.09
pH	6.0-9.0	6.0-9.0

The regulation educates on how these numbers were calculated. Because of a lack of industry-wide information regarding photography and its manufacturing process, a Best Practicable Technology standard was advisable [26] Regulators were also unable to effectively sub-categorize the manufacturing process because of this lack of information. It is thus more advisable to use a less stringent standard (BPT) and to allow more discretion in the permit process when comprehensive data is unattainable. To mandate stricter requirements without an understanding on its economic impact would offend the overarching purpose of technology-based standards, which seeks to balance the interests of the environment with the economic interests of the regulated industry.

b.) Best Available Technology

Best Available Technology Economically Achievable (BAT) mandates "the maximum feasible pollution reduction for an industry."[27] A more stringent standard than BPT, this approach designs regulation on "the optimally operating plant," seeking to maximize pollution reduction without causing large-scale facility closures, even if the technology is not widely used in the industry.[28] The role for innovation is even more relevant in this circumstance. While this standard only requires what is theoretically achievable by the industry, there is always pressure to find a more cost-efficient means of compliance. Scientists and engineers undoubtedly play a major part in the research and development of these cleaner technologies. Similarly, when determining BAT, regulators must consult with those developing technology in a specific industry and determine the outer limits of pollution reduction.

As stated above, Best Available Technology Economically Achievable (BAT) is most applicable in a circumstance where the regulated contaminants are especially harmful, or when the costs of a "heightened" technology can be reasonably borne by the industry. In

---

[26] *Id.* § 459.12(b).
[27] *Id.* at 45.
[28] GROSS, *supra* note 22, at 41 (internal quotations omitted).

1976, EPA codified regulations regarding the manufacturing process of oil-based paints.[29] These regulations apply a BAT standard by simply declaring that "there shall be no discharge of process waste water pollutants to navigable waters."[30] Regulators determined that the state of technology permitted a wholesale ban on the release of oil-based paint into the nation's water. Correlatively, such an approach reflects the danger that oil-based paint poses to the nation's water supply. Also, from the perspective of a layperson, it follows logic to place the most restrictive regulation on something like paint. In other words, it would be unsettling to think that manufacturers could lawfully release paint into our water supply.

B. The Role of Scientists in Protecting Water Quality

Congress enacted the CWA to "restore and maintain the chemical, physical and biological integrity of the nation's waters."[31] The goal of the CWA is to eliminate "the discharge of pollutants into the navigable waters," and in the interim, to attain "water quality which provides for the protection and propagation of fish, shellfish, and wildlife and provides for recreation in and on the water."[32] The passage of the CWA "marked the ascendancy of water-quality control to the status of a major national priority."[33]

Scientists – biologists, botanists, lymnologists, arborists, zoologists – play a key role in protecting waters for which technology based standards alone are not sufficient to protect water quality and ensure that otherwise allowable pollutant loadings do not exceed the assimilative capacity of the water receiving the discharge. In modern times, this is largely accomplished by developing and implementing "water quality standards" under the Clean Water Act. The Clean Water Act is intended to ensure protection of our nation's waters by requiring states to take certain measures to ensure both that safe and healthy waters would not become more polluted and that impaired waters would be restored. To meet these objectives, the CWA requires EPA to set water quality "criteria" to protect uses, such as fishing, swimming, drinking and recreation. The CWA then has states identify impaired waters, establish pollutant load limitations for such impaired waters to ensure that they meet water quality standards, and develop an implementation plan to see to it that such standards are met for all waters in a state.

To achieve these ends, section 303 of the CWA requires the establishment and implementation of water quality standards.[34] States are required to establish water quality standards subject to review and approval by EPA.[35] The Supreme Court has described the achievement of water quality standards as one of the CWA's "central objectives."[36]

The linchpin to achieving water quality standards is the "Total Maximum Daily Load" ("TMDL") program of Section 303(d) of the Clean Water Act.[37] This provision establishes a

---

[29] 40 C.F.R § 446 (2008).
[30] *Id.* § 446.13.
[31] 33 U.S.C. § 1251(a) (2008).
[32] *Id.* § 1251(a)(1) and (2).
[33] Monongahela Power Co. v. Marsh, 809 F.2d 41, 45-46 (D.C. Cir. 1987).
[34] 33 U.S.C. § 1313 (2008).
[35] *Id.* § 1313(a).
[36] Arkansas v. Oklahoma, 503 U.S. 91, 105 (1992).
[37] 33 U.S.C. § 1313(d).

detailed interrelated process. First, section 303(d)(1) of the CWA requires every state in the mid-Atlantic to identify every segment of the waters within its boundaries that do not meet or are not expected to meet applicable water standards even after the imposition of best-practicable technology-based effluent limitations, secondary treatment standards for publicly owned treatment works, and controls on thermal discharges.[38] EPA refers to these impaired waters as "Water Quality Limited Segments" ("WQLSs").[39]

Following its identification of impaired waters, states must determine the maximum tolerable pollution so that pollutant loading in an impaired water body does not, taking into account seasonal variations and allowing an ample margin of safety, exceed the standards established for the water body. Section 303(d)(1)(C) requires states to develop the limits or load.[40] In its simplest term, a total maximum daily load is a calculation of the maximum amount of a pollutant that a waterbody can receive and still meet water quality standards, and an allocation of that amount to the point and nonpoint sources of that pollutant.

Once states complete the above steps, they are to submit the list of standards and allowable loadings to the United States Environmental Protection Agency ("EPA") for review. Section 303(d)(2) states: "Each State shall submit to the Administrator ... for his approval the waters identified and the loads established under paragraphs (1)(A), (1)(B), (1)(C), and (1)(D) of this subsection."[41] EPA either must approve or disapprove the list of impaired waters and allowable loadings. If EPA disapproves, EPA must promulgate a new list of standards or allowable loadings within 30 days.[42]

Thus, the CWA establishes a dual approach to achieving water quality goals, with national technology based standards as the primary tool, and local, ambient-based standards as a safety net for waters for which technology based standards alone are not sufficient to achieve water quality standards.[43] Scientists in turn play an important role at every stage of this process. They help develop criteria, set standards, identify impaired waters, develop assimilative loading capacities, and determine allowable discharges. They also help to design and enforce permits that incorporate technology based and water quality based standards.

Bringing it All Together: Science, Engineering and Permitting

Permits serve "to transform generally applicable effluent limitations and other standards—including those based on water quality—into the obligations (including a timetable for compliance) of the individual discharger."[44] A century ago, Congress sought to hinder the "[d]eposit of refuse in navigable waters"[45] by enacting The 1899 Rivers and Harbors Act, a

---

[38] *Id.* § 1313(d)(1)(A) and (B).

[39] 40 C.F.R. § 130.7(b)(1) and (2) (2008).

[40] 33 U.S.C. § 1313(b)(1)(C).

[41] 33 U.S.C. § 1313(d)(2).

[42] *Id.*

[43] *See* James R. May, *The Rise and Repose of Assimilation-Based Water Quality, Part I: TMDL Litigation*, 34 Envt'l L. Rep. 10247 (Env. L. Inst. 2004).

[44] EPA v. California ex rel. State Water Resources Control Board, 126 U.S. 200, 204-205 (1976).

[45] 33 U.S.C.A. § 407 (2008).

"rudimentary permitting system" that was the precursor to the modern-day Clean Water Act.[46] In subsequent legislation, the federal government attempted to devolve water regulatory efforts to state governments, with federal authorities taking more of an oversight role.[47]

Under the Clean Water Act the federal government established nationwide system of limiting water pollution. This permit process, known as the National Pollutant Discharge Elimination System (NPDES), is primarily enforced by state governments, although regional Environmental Protection Agency offices administer the process when state agencies lack the relevant authorization.[48] While water pollution occurs in a variety of forms, the NPDES system focuses primarily on industrial and municipal facilities that release pollutants into surface waters.[49]

At any point in the evolution of permitting systems, government relied on scientists and engineers to set effective standards that simultaneously protected water resources and enabled industry to operate without unreasonable barriers. Many questions arise during the process. Are some pollutants more destructive than others? Can some pollutants exist in water without effecting the health and safety of the public? Where should regulators draw the line? These questions fall well outside the realm of common knowledge. In determining the criteria for a meaningful permit process, government relies on the scientific and engineering communities.

Similarly, when industrial and municipal facilities apply for permits, scientists and engineers play a central role in gathering the relevant information and communicating with their counterparts in government. For example, NPDES applications require extensive quantitative data about the structure of a facility and the contents of its discharges.[50] This information is so esoteric that companies sometimes hold "'signing ceremonies'" where corporate officers have an opportunity to question those responsible for the content of the application.[51] Even when a facility receives a permit, technical assistance is still required to decipher the permit's terms, conditions, and methods of calculation.[52] Without this assistance, a company or municipality will not be able effectively to suggest revisions or otherwise comment on the permit process. Furthermore, NPDES permits require a facility to records its own discharge levels and submit reports to the relevant government authority.[53] Clearly, each step in the process calls for technical assistance.

Scientists, Engineers and Water Rights

For centuries, people have fought over access and control of water. Whether in 18th century England or the modern day American West, water rights are broad and easily

---

[46] JOEL M. GROSS & LYNN DODGE, *supra* note 21 , at 5

[47] *Id.* at 6.

[48] Environmental Protection Agency, National Pollutant Discharge Elimination System (NPDES), *available at* http://cfpub.epa.gov/npdes/.

[49] *Id.*

[50] LYNN M. GALLAGHER, *supra* note 22, at 23-24.

[51] *Id.* at 25.

[52] *Id.* at 27.

[53] GROSS, *supra* note 21, at 33-34.

offended.[54] When adjudicating disputes over water rights, the importance of expert testimony is unquestioned. Often, litigation involves private parties seeking to use water, a public resource, for private purposes.[55] Water appearing on Earth's surface is an overwhelmingly small percentage of our potable water.[56] In other words, most of the water consumed by humans (and subject to potential litigation) exists underground. Engineers, geologists, and other scientists are therefore essential in pinpointing water sources and explaining their potential use. Without this technical knowledge, settlement of disputes in water cases would extremely difficult and potentially very costly, both in terms of dollars and water misuse.

Epilogue and Conclusion

Engineers and scientists will continue to play a featured role in water regulation for the next 100 years. While water quality has improved, according to EPA, approximately 60 percent of assessed waters nationally are safe enough for fishing and swimming. Wetland losses have slowed to one-quarter the rate of 30 years ago. These efforts have helped clean up pollutant discharges throughout the nation, from waters of the Delaware Basin and the Inland Bays, to the Baltimore Harbor and the Chesapeake Bay, to the Susquehenna and the Three Rivers system, to Lake Erie, to the Potomac and the Anacostia, to the Blackwater, to name a few.

Notwithstanding what's been done, to borrow from David Frost, it seems we still have far to go before we sleep. Pollutant discharges from factories are on the upswing for the first time in 30 years. Polluted runoff from farm fields, city streets and parking lots is virtually uncontrolled, choking half the nation's waters. Wetlands continue to be destroyed at an alarming rate, nearly 250,000 acres (one-half size of Delaware) each year. Thousands of lakes, streams, and miles of rivers used by 3 in 4 in the mid-Atlantic fail to meet some basic water quality standard.

From 1998 to 2000, the percentage of polluted rivers rose from 35 to 40 percent, shorelines from 12 to 15 percent, and polluted estuaries – the best measure of ecosystem health -- from 44 to 51. The latest statistics reveal there were more than 400 beach closings because of health advisories, 2,500 waters with fish consumption advisories or bans, 37 "water outbreaks" in 17 states assessed by the CDC, including red tide and algae outbreaks in Delaware, Physteria in Maryland and Delaware, and shellfish contamination in the Chesapeake.

These statistics obscure that the vast majority of the nation's waters are not even assessed. The Bush Administration has refused to issue a report to Congress on the state of the nation's waters. The National Academy of Sciences recently concluded there is too little data to know much of anything about water quality, and less being developed recently. States are doing less, amid political and economic shortfalls. The Heinz Fund picked up the slack by releasing its Environmental Indicators Project about five years ago. It suggests water quality is

---

[54] Raphael J. Moses, *The Expert Engineer: The Water Lawyer's Best Friend*, available in WATER LAW: TRENDS, POLICIES AND PRACTICE 11 (Kathleen Marion Carr & James D. Crammond ed.) (American Bar Association 1995).

[55] *Id.* at 12-13.

[56] U.S. Geological Survey, Where is the Earth's water located?, *available at* http://ga.water.usgs.gov/edu/earthwherewater.html

worse now than 30 years ago, but due to the lack of data, concluded that no trends could be forecasted.

Looking forward, the regulation of water requires collaboration between the law, science and engineering. Healthy communication between government and science and engineering is critical to safeguard the public's overwhelming interest in safe, clean and accessible water.

Appendix

I. SCIENCE, ENGINEERING AND WATER POLLUTION SOURCES, TYPES AND CONTROL TECHNOLOGIES

Background: Driving Forces of Treatment Technology Development

(a) government standards
(b) chemical/physical nature of pollution
(c) cost effectiveness
(d) potential consideration

A. **Types and Sources**

1. *Water Quality Parameters of General Interest*

Fecal coliform	Taste & order	Phosphorous
TSS	Hardness	Nitrogen
Color	Alkalinity	
Hardness	TDS	
	Iron & Manganese	

2. *Typical Parameters*

BODs	TSS	Metals
COD	pH	toxics
O&G	Nutrients	

3. *Important Terms*:

a. Completely soluble or miscible H_2O & Alcohol
b. Suspension & sediment
c. Emulsion (oil & H_2O)

B. **Technologies**

1. *Coagulation & Sediment*

a. Usually preceded by precipitation hydroxide or chromium hydroxide, ferric iron.
b. Opposite method is dissolved air floatation (DAF)

2. *Biological Treatment*

a. Aerobic
b. Anerobic

3. *Chemical Oxidation*

a. Example: Chlorine treatment for disinfection ozonation (like aeration chemical).

4. *Membrane Processes*

a. Gatekeepers - only allow certain molecules to pass.
b. Generally made from organic polymers - reverse osmosis; ultrafiltration, electrodialysis.
c. Typically used with desalination plants.

5. *Ion Exchange*

a. Water softener.
b. Fills up "exchange sites in H_2O that would otherwise be filled by calcium/magnesium in hard H_2O .

6. *Sorption Processes*

a. Example: fish tank.
b. Activated Carbon for removal of organic compounds, or for polishing and toxics.

7. *Vapor Phase Systems*

a. Frowned upon - Wastewater problem = air pollution.

8. *Polishing Systems*

a. For solids
b. Filtration
c. Settling
d. Marshes (natural/manmade)

9. *Subsurface considerations* (bioremediation)

a. Pumping
b. In site treatment (proprietary bacteria)

C. **Wastewater Technologies**

1. *Primary*
a. Protect equipment in the treatment process
b. Screening – rags, large material
c. Pretreatment – typically to reduce odor by adding chemicals to wastewater
d. More obnoxious materials removed
e. 1 – 2% of the pollutants removed

f. Costs – 5 – 10% of plant costs
g. $100,000 – $200,000/million gallons of capacity
h. Significant source of odor release

2. *Removes*
   a. 40 – 50% of solids
   b. 20 – 30% of organics
   c. Small reduction in toxics unless chemicals added (chemically enchanced primary – CAP)

3. *Physical Processes*
   a. Settling – larger, heavier materials settle to bottom of basin and removed. (Primary Sludge)
   b. Lightest material float to top of the basins and are scrapped off. (Scum)

4. *Settling Basins*
   a. May be circular or rectangular and may be stacked on congested sites

5. *Costs*
   a. $1.0 – 1.5 million per million gallons of capacity
   b. Significant odor release possible; may require covering of basins and treatment of removed gases.

6. *CAPS* (chemicals added to):
   a. Improve settling – up to 80% removal
   b. Remove nutrients
   c. Improve toxics removal
   d. Significantly increases volume of primary sludge
   e. Cannot meet secondary treatment standards but may be adequate technology

7. *Secondary Treatment*
   a. Typically follows primary treatment
   b. Designed to remove:
      1) Solids that didn't settle
      2) Floatables that don't float
      3) Organic material dissolved in the water
      4) 85 – 90% removal of solids
      5) 85 – 90% removal of organics
      6) 50 – 90% removal of most toxics

8. *Biological Process*
   a. Soil bacteria are grown in large quantities (called activated sludge or biological films) the bacteria consume remaining pollutants, the bacteria are then removed by secondary settling.

9. *Settled Bacteria*
   a. Some returned to aeration to consume more waste.
   b. Some thrown away (Secondary Sludge)

10. *Various technologies*
   a. Activated sludge
   b. Biological towers
   c. Pure oxygen
   d. Because the process is biological, it's more difficult to control.
   e. Costs – $2.0 – 3.0 million per million gallons of capacity including primary treatment.

11. *Air emissions*
   a. Can release odors but usually reduced because secondary uses aerobic process.
   b. Can release substantial quantities of VOC's and may release air toxics
   c. Some reduced efficiency in cold climates.

12. *Advanced Treatment*
   a. Treatment added to primary and secondary to get even higher levels of removal
   b. 95 – 98% removal of solids, organic matter
   c. 80 – 95% removal of toxics

13. *Other technologies*
   a. Filters – to improve solids/organics

   b. Chemical addition
      1) Lime, ferric chloride, alum, polymers to remove
      2) Phosphorus or improve toxic removal

   c. Biological – grow bacteria that convert ammonia to nitrogen gas, hence removing nitrogen.

   d. Polishing
      1) Membranes
         a) Very high quality removal by technology (including only certain chemicals to pass through a membrane – Reverse Osmosis).
         b) Used for highest quality water, or removal of toxics or desalination.
         c) Has been limited to smaller plants because of costs.
         d) Technological improvements are making it more attractive for larger plants.

   e. Electrodialysis Reversal
      1) Electrical charges are used to remove undesirable materials or to separate materials.
      2) Use heavily in desalination.
      3) Costs – $3.0 – 7.0 million per million gallons of capacity including the costs of primary and secondary operating costs.
      4) Operating costs are 2 – 3 times those of secondary

f. Air Emission

1) Generally not of concern if preceded by secondary facilities.

g. Disinfection (usually drinking water technologies)

1) Follows other treatment units
2) Designed to kill harmful bacteria and inactivate viruses
3) Chlorine
    a) Extensively used until recently
    b) Chlorine by-products have been found to be carcinogenic
    c) If chlorine used, dechlorination may be used to remove by-products. (Add another chemical – sulfur dioxide)

4) Ozone
    a) Used instead of chlorine where effluent quality is high.
    b) Avoids by-products

5) Ultra-violet
    a) Used on high quality effluent.
    b) Becoming more affordable.
    c) Efficiency affected by solids

14. *Management of Residuals*

a. Wastewater's "forgotten sister"

1) Became a major national issue when secondary treatment was imposed because of larger quantity solids.

b. Broadly not treated as a solid waste.

c. Screenings (very obnoxious)

1) Almost universally buried in landfills, sometimes burned

d. Grit (similar to screenings)

e. Scum

1) Can be dewatered and landfilled.
2) Can be burned because has high energy value.
3) Can be digested with primary/secondary sludge.

f. Primary/Secondary Sludge

1) Contains large quantities of water
2) Contains large percentage of putrescible solids
3) Major odor source
4) Primary/Secondary Sludge Technology
    a) <u>Stabilization</u> – to reduce the putrescible nature
    b) <u>Dewatering</u> – to reduce the amount of water
    c) Disposal by:
        i. Beneficial reuse

ii. Incineration
iii. Landfilling
iv. Disposal at sea is no longer permitted in the U.S.

g. Stabilization

1) Chemical addition, usually lime, doesn't destroy putrescible solids; merely controls their decomposition
2) Aerobic or anaerobic digestion (with or without air) bacteria are used to reduce putrescible solids by 60 – 70%
3) Anaerobic digestion produces useable methane gas

h. Dewatering

1) Design to convert sludge that is 92 – 97% water to sludge that is 60 – 75% water.
2) May use chemicals to aid dewatering.
3) Technology to remove water.
a) Centrifuges
b) Belt presses
c) Filter presses

i. Disposal

1) Sludge has beneficial value as a soil conditioner/fertilizer if properly prepared.
2) Reuse technologies include:

a) Agricultural application – sludge is applied to agricultural land in liquid or dewatered state.

b) Composting – various equipment is used to further stabilize and dry sludge until it reaches a humus like consistency. It then can be applied to agricultural land and used in landscaping.

c) Pelletizing – dewatered sludge is thoroughly dried and broken into small, fertilizer like pellets for reuse on agricultural land and landscaping.

d) Incineration – dewatered sludge is burned at high temperatures. If the sludge is dry enough, incineration occurs without adding fuel. Heat/steam are often recovered from the incineration. The high temperatures substantially reduce but do not eliminate air emissions concerns.

e) Landfilling – dewatered sludge is burned in an approved landfill. The sludge is stabilized by digestion or by chemical addition prior to landfilling.

f) Costs

i. Landfilling – generally the least expensive but very difficult to cite – $80 – 200 per ton

ii. Agricultural Application – \$100 – 300/ton.

iii. Composting, Incineration Drying – \$200 – 800/ton

Note: In colder climates, beneficial reuse options may require storage for several months each year.

j. Technological Advancement

1) Basic Technology – settling and biological treatment have changed very little in the last 100 years

a) Improvements:

i. Equipment – more energy efficient
ii. Controls – more intelligent systems with expanded computer use
iii. Chemicals – more variety, especially polymers
iv. Success – still achieved by the men and women who operate and maintain every hour of every day

II. TECHNOLOGY, SCIENCE AND DRINKING WATER

A. **Contamination**

1. Physical
   a. Solids
   b. Color

2. Biological
   a. Bacterial
   b. Viral
   c. Giardia

3. Chemical
   a. Toxic Metals
   b. Organics
   c. Special Concerns – Lead

B. **Standards – Basis – Risk Profile**

1. *Management versus Control*
   a. watershed protection
   b. Aquifer management
   c. Corrosion control

2. *Control (Treatment)*
   a. Pretreatment
   b. Chemical addition
   c. Chlorine
   d. Ozone
   e. Coagulation/Settling

f. Filtration
g. Disinfection
h. Ammonia Control
i. Storage (Covered)

A Century of Transport – A Personal Tour

Stuart W. Churchill
The University of Pennsylvania

The industrial chemists and mechanical engineers who banded together in 1908 to form the AIChE were self-taught experts in the design and operation of equipment to carry out chemical processing. Although graduates with a degree in chemical engineering had been entering industry for over a decade, none of them were among the founding members because they failed to meet the qualification of "years of responsible charge". The skills possessed by the founding members can be discerned from the early identification of chemical engineering with what we now call the *unit operations*. Remarkably, these unit operations were all physical and did not invoke chemical reactions explicitly.

The work of the first chemical engineers could be distinguished from that of chemists, other than perhaps "industrial" ones, by their utilization of continuous processing and thereby of an Eulerian rather than a Lagrangian framework for modeling, as well as by their involvement with transport, in particular with fluid flow, heat transfer, and mass transfer. Their work could be distinguished from that of mechanical engineers by virtue of their involvement with chemical conversions and chemical separations, and over the century with an increasing variety of materials, many of them synthetic and new to the world. Accordingly, the differences in the approach of chemical engineers and mechanical engineers to fluid mechanics, heat transfer, and thermodynamics have continued to widen.

The joint designation of fluid mechanics, heat transfer, and mass transfer as *transport* rather than as *unit operations* and/or as separate topics, began with the publication in 1960 of the most influential book in the history of chemical engineering, namely *Transport Phenomena* by Bob Bird, Warren Stewart, and Ed Lightfoot. Most advances come at a price, and in the long run this consolidation and theoretical upgrading has resulted in a lesser preparation of our students in applied fluid mechanics and heat transfer *per se*.

Although our understanding of transport has greatly progressed over the century, this subject now has lesser role in education and practice, because the total number of technical courses in the engineering curriculum has regularly been decreased, and because additions of new subject matter have led to a reduction in those devoted to engineering science. These "advances" have come at the price of a poorer preparation for our students in transport itself. The current "in" topics of biotechnology and nanotechnology involve transport but mostly at such a smaller scale that what I will be describing is applicable, if at all, only as a guideline.

My objective today is to review the evolution of the skills and practices of chemical engineers in dealing with transport while avoiding, insofar as possible, those aspects being addressed by the other speakers, namely the validity of the Navier-Stokes equations, the role of numerical computation, and biological applications.

Because of the aforementioned expansion of knowledge and applications in transport, its breadth, even in a classical sense, is such that I cannot cover all aspects or even the evolution of any one aspect completely. My coverage today is therefore both highly selective and very arbitrary. My initial plan was to trace developments in 1) turbulent flow and convection in tubular flow, because much chemical processing occurs in that regime and geometry, 2) transport in packed and fluidized beds, because the advances have been made almost exclusively by chemical engineers, 3) mass transfer, because chemical engineers have an almost exclusive role therein, 4) non-Newtonian flow, because chemical engineers have a large although not exclusive role therein, and 5) combustion, because chemical engineers appear to have the most appropriate background.

However, this seemingly logical approach proved to be awkward to implement, and after a struggle with it, I switched to the evolution and adaptation of critical concepts and methodologies, namely the identification of compound variables, irrespective of the process itself, and the development of predictive expressions, correlating equations, and analogies. The path of progress itself is traced for only two topics in transport, namely turbulent flow and turbulent convection, and even then only in terms of a few highlights. My own primary field of research, namely combustion, fell by the wayside. I conclude with some general observations concerning the influence of AIChE programming at its meetings, and of the sources and extent of funding for advances in transport.

The date of 1908 is a signal one for us here today, but it is not a dividing line in terms of chemical technology. Many of the concepts that now characterize the contributions of chemical engineers to society originated earlier. Also, as painful as it is to admit, many of our treasured and most useful concepts originated, even after 1908, with non-chemical engineers. Rather than be chauvinistic, I have chosen to identify the actual origins of these concepts and to relate our subsequent applications and improvements to those forerunners and professional associates, even though they did not have the wisdom or foresight to identify themselves as chemical engineers.

My choice of and order of individual topics is not important because they are intended merely as illustrations. I concede and apologize that several of the choices were influenced by my own involvement – not because of that involvement but because I am more confidence in and ready access to the historical record in those instances. In defense of that position, I may cite my presence on the sidelines or in the arena of chemical engineering for 88 of the last hundred years, my commitment to chemical engineering as a profession for 70 of them, my formal association with the AIChE for 69, and my focus on transport over most of that span.. Of course these experiences do not qualify me to predict the future direction of transport, so I will stick to my assigned topic – a review and interpretation of progress during *The First Century of the AIChE* and leave the predictions to others. .

I am taking one liberty that may run contrary to the mood of this celebration of our roots in transport, namely citations of wrong turnings. Such identifications are not meant to impugn those who originated the false or inferior concepts and expressions, because all concepts in science and engineering are eventually discarded or improved upon; that is the process by which progress is made. Indeed, the original concepts and expressions that are later proven wrong are in most instances a necessary first step, and their discard usually reflects the acquisition of new information or a new insight.

I have taken one other liberty, namely favoring technical continuity over strict chronology. The resulting order of topics might be confusing to undergraduate students but not to this audience because the topics taken out of temporal sequence are familiar ones.

Continuity and Conservation

The equations of conservation – the Navier-Stokes equations and their counterpoints for energy and species – are the starting point for most theoretical work on transport, but, out of deference to Howard Brenner, I will not trace the development of these equations nor examine their validity except to mention one contrary opinion that brought me up short at the time. The renowned physicist George E. Uhlenbeck, whom I was fortunate enough to have as a mentor, was so frustrated by his failure to confirm or disprove the N-S equations by reference to statistical mechanics, which he considered to be a better starting point, once wrote "Quantitatively, some of the predictions from these equations surely deviate from experiment, but the *very* remarkable fact remains *that qualitatively* the Navier-Stokes equations *always* describe physical phenomena sensibly.....The mathematical reason for this virtue of the Navier-Stokes equations is completely mysterious to me."

Conceptual and Compound Variables

The use of conceptual and compound variables in chemical engineering is so pervasive that we may forget that they are arbitrary, that in many instances their introduction was a consequence of great ingenuity, and that we almost blindly place our trust in the unexplored variables buried within the compound ones. I will call your attention to a few of them.

The heat transfer coefficient The heat transfer coefficient is perhaps the best example of a revolutionary concept and of a compound variable that has become imbedded in chemical engineering. Isaac Newton in 1701 noted that the rate of heat transfer to a surface in free convection was proportional to the area times the difference between the free-stream temperature and the temperature of the surface. He thereby inspired the replacement of four variables with one, namely the coefficient of proportionality that we now call the *heat transfer coefficient*. This concept, although usually encompassing a degree of approximation and scorned by some, has remained in active use for over 300 years. Eventually it was extended to forced convection in tubular flow by replacing the free-stream temperature with the mixed-mean temperature of the fluid, and was adapted for mass transfer in terms of fugacities, partial pressures, and/or concentrations. The friction factor, drag coefficient, and orifice coefficient are closely related conceptual variables for momentum transfer. These five coefficients have been utilized by chemical engineers for more than one hundred years, and they remain invaluable and irreplaceable. Most of our technical data-base is compiled in this form.

The equivalent thickness for pure conduction This quantity, as defined by $\delta_e = = j_w/k\Delta T$ (or *h/k*) has repeatedly been proposed as an alternative to the heat transfer coefficient, but, with one notable exception, it has always proven inferior in terms of correlation, generalization, and insight. That exception was its use by Langmuir in 1912, in the context of an analysis of heat losses by free convection from the filament of a partially evacuated electrical light bulb. He utilized the concept of an equivalent thickness for thermal conduction to derive an approximate expression for the effect of the curvature of the surface of the wire, *vis-à-vis* a vertical flat plate, on the rate of heat transfer by free convection. The resulting expression, which, after nearly a century, has not been improved upon and which is based on the applicability of the log-mean area for conduction across a cylindrical layer, is

$$Nu = 2/\ln\{1 + (2/Nu_f)$$

Here Nu_f is a correlating equation for free convection from a vertical flat plate. It should be noted that the effective thickness has vanished from the final expression. This relationship has been found to be uniquely useful as an approximation for both free and forced convection from a horizontal cylinder as the Grashof and Reynolds numbers, respectively, approach zero, and is readily adapted for the region of the entrance in laminar tubular flow and for mass transfer. When applied to a spherical layer this concept invokes the geometric-mean area and leads to an exact asymptote for a decreasing Reynolds number or Grashof number. "*Vive la exception*" due to Langmuir.

The mixed-mean velocity The mixed-mean velocity has proven to be a very useful concept in chemical engineering because of the pervasive use of tubular flow in chemical processing. For constant density it is defined as

$$u_m \equiv 2\int_0^1 u_r\left(\frac{r}{a}\right)d\left(\frac{r}{a}\right)$$

For fully developed flow, the mixed-mean velocity is simply equal to v/A_x, where v is the volumetric rate of flow and A_x is the cross-sectional area. The integration is thereby rendered unnecessary in so far as v is known.

The mixed-mean temperature The mixed-mean temperature has also proven to be a very useful quantity in chemical engineering for the same reason. For a fluid of constant density and heat capacity in a round tube it is defined as

$$T_m \equiv 2\int_0^1 T_r\left(\frac{u_r}{u_m}\right)\left(\frac{r}{a}\right)d\left(\frac{r}{a}\right)$$

There is, however, no directly and simply measurable quantity for the mixed-mean temperature analogous to v/A_x for the mixed-mean velocity. I will later describe one instance in which this concept has led us astray for nearly a century, but that exception does not undermine its overall usefulness.

Mixed-means in general The corresponding expression for the mixed-mean concentration can readily be inferred, and the slightly more complex expressions for the mixed-mean velocity, temperature, and concentration in the cases of variable density, heat capacity, and/or concentration can readily be formulated, but they are less useful because of their relative complexity and the difficulty posed by their measurement.

Fully developed flow The velocity profile develops down-stream from the entrance to a tube and eventually approaches an asymptotic one in both the laminar and turbulent regimes. The local value of the friction factor decreases and approaches an asymptotic value in laminar flow and in turbulent flow in a rough but not a smooth pipe. This asymptotic state for the velocity distribution is classified as *fully developed flow*. The rate of development depends on the geometrical configuration and the Reynolds number, and in the turbulent regime the asymptotic velocity distribution itself also depends on the Reynolds number. In tubes with a large ratio of length to diameter, the concept of fully developed flow is often utilized as an approximation for the entire length of the tube. Strictly speaking, fully developed flow implies constant density and viscosity.

The friction factor for artificially roughened tubes In 1932 Nikuradse, a student of Prandtl, confirmed experimentally beyond all question that the friction factor for turbulent flow in a smooth pipe decreases indefinitely as the Reynolds number increases. In 1933 he discovered that if an artificial uniform roughness is imposed on the surface of the pipe the friction factor instead approaches an asymptotic value. He devised a correlating equation for this asymptotic value as a function of the amplitude of the roughness divided by the diameter. Although the results of this experiment and analysis were of no direct practical value because the roughness in commercial piping is anything but uniform, this work of Nikuradse not only provided insight but was an essential precursor for the practical developments that followed.

The friction factor for commercial (naturally roughened) tubes Colebrook in 1938-39 reported on his measurements of the friction factor for real piping of various types, and proposed a truly ingenious concept for the representation of their "effective" roughness, namely the value that produced the same asymptotic friction factor as that given by the aforementioned correlating equation of Nikuradse for uniform roughness. The practical utilization of this concept requires a tabulation of effective values of the roughness for various types of piping, such as steel, both new and aged and glass. (As an aside, the tables of roughness in most of our current textbooks and handbooks are badly out of date and should be revised to reflect modern processes for the manufacture of tubing.) Colebrook also formulated a generalized empirical equation for the approach to that asymptotic value as the Reynolds number increases. The turbulent regime of every friction factor plot that you have ever utilized is merely a graphical representation of that equation.

The equivalent length A similar concept to that of the effective roughness has long been utilized to estimate the pressure drop due to pipe fittings such as valves and elbows, namely the length of straight smooth pipe that results in approximately the same pressure drop for a given rate of flow. This concept implies that the dependence of the pressure drop on the rate of flow is the same for the fitting as for straight piping, which is a crude approximation at best. The application of this concept for design requires a table of values. Again, the tables in our handbooks are out-of-date and of unknown reliability.

"Plug flow" An unfortunate concept pervades much of the literature of chemical engineering, and in particular that of reactor engineering, namely "plug flow", which occurs physically only when some semi-soft solid such as ice cream is pushed through a tube by a plunger. Turbulent flow does not approach "plug flow" as the Reynolds number increases; the velocity still goes to zero at the wall and the velocity profile is better characterized as parabolic-like than as flat. The idealization of "plug flow" does result in simple and reasonably accurate solutions for the chemical conversion for some conditions, but also very erroneous ones for others. The approximate solutions for chemical conversions that are based on "plug flow" need not be discarded; they can simply be reinterpreted as those for a physically conceivable condition, namely perfect radial mixing, that is for $Pr \rightarrow 0$ and $Sc \rightarrow 0$. Solutions for "plug flow" appear in the literature of heat transfer, but only in the context of lower bounds for Nu.

Integral boundary-layer theory When I was a graduate student, the advanced textbooks on fluid mechanics and heat transfer included a section on integral-boundary-layer theory. This methodology, which was apparently devised independently by von Kármán and E. Pohlhausen in 1921, was based, for flow, on the postulate of an arbitrary velocity distribution, thus allowing an analytical integration of the momentum balance to obtain a closed-form expression for the drag coefficient. The equivalent process was applied for the temperature distribution to obtain an expression for the heat transfer coefficient. One positive measure of progress in the field of transport is the disappearance of this concept from modern books on fluid mechanics and heat transfer.

Potential (inviscid) flow and the boundary layer concept Boussinesq in 1905, and thus just three years before the founding of the AIChE, derived solutions for forced convection in inviscid flow over immersed bodies. These solutions are now recognized to have practical value only as asymptotes for $Pr \rightarrow 0$ and in the boundary layer model of Prandtl, in which the latter conceived of a thin layer of slowly moving fluid next to the surface and inside a regime of inviscid flow. This concept remains alive today, but its limitations should not be overlooked.

Free streamlines The boundary between moving and non-moving segments of a fluid stream is called a *free-streamline*. The pressure is constant along such a line or surface. This concept is an important one in aerospace engineering and in civil engineering, and has at least one lasting application in chemical engineering, namely theprediction of a value of 0.5 for the coefficient for both a planar and circular Borda entrance, and a value of $\pi/(2+\pi)$ = 0.6110... for both a planar and a circular orifice. These values are, however, only approximations. For example, the actual limiting value for a sharp-edged orifice in a round tube is 0.5793 rather than 0.6110.

Criteria for turbulent flow In 1883 Osborne Reynolds, on the basis of the most famous experiment in the history of fluid mechanics, deduced that below a certain numerical value of the dimensionless group $Du_m\rho/\mu$, now named for him, a perturbation introduced into a fluid flowing through a pipe would dampen out. That value is generally taken to be 2100. However, the value below which the flow is totally free of turbulent fluctuations, namely $Re \cong 1600$ or $a^+ = a(\tau_w \rho)^{1/2}/\mu = Re(f/8)^{1/2} \cong 56.6$, and that above which the flow is fully turbulent, namely $Re \cong 4020$ or $a^+ = Re(f/8)^{1/2} \cong 150$, are of more practical utility in terms of transport.

Fully developed convection The concept of fully developed convection is much more subtle and its initial formulation required some ingenuity. The temperature of the fluid stream in a heated or tube increases continuously but, by virtue of the nearly linear dependence of the enthalpy on temperature, a dimensionless temperature that approaches an asymptotic value may be defined for some particular thermal boundary conditions.

For example, if a uniform heat flux density is imposed on the fluid at the wall, the mixed-mean temperature of the fluid must increase linearly with axial distance insofar as the density and heat capacity can be considered invariant. It follows that the temperature of the wall must thereafter also increase linearly at the same rate, and that $(T-T_0)/(T_m-T_0)$, $(T_w-T)/(T_w-T_m)$, and the heat transfer coefficient must approach asymptotic values. Fully developed convection for uniform heating is thus defined by the near-attainment of asymptotic values for $(T-T_0)/(T_m-T_0)$, $(T_w-T)/(T_w-T_m)$ and the heat transfer coefficient. It follows that $\partial T/\partial x \to \partial T_w/\partial x \to \partial T_m/\partial x \equiv dT_m/dx$, which allows simplification of the differential energy balance.

A uniform wall-temperature also results in an approach to an asymptotic value for the heat transfer coefficient and for $(T_w-T)/(T_w-T_m)$ but not for $(T-T_0)/(T_m-T_0)$,. Seban and Shimazaki in 1951 recognized that the attainment of an asymptotic value of the first of these quantities implied that its derivative could be equated to zero. They thereby derived

$$\frac{\partial T}{\partial x} = \frac{\partial T_m}{\partial x}\left(\frac{T_w - T}{T_w - T_m}\right)$$

The two terms on the right-hand side are both independent of the radius, and therefore $\partial T/\partial x$ can be replaced by dT/dx. Substitution of this expression in the differential energy balance results in some simplification.

Fully developed convection is often utilized as an approximation for the entire length of the tube. Most analytical and numerical solutions for thermal convection are for fully developed convection and thereby for uniform heating or uniform wall-temperature because of the relative simplicity of the behavior. The former condition can be approximated in practice by electrical-resistance heating of the tube wall or by equal counter-enthalpic flow, and the latter condition by subjecting the outer surface of the tube to a boiling fluid (or for cooling to a condensing fluid). Constant viscosity, as well as density and heat capacity, are implied by the concept of fully developed thermal convection. The concept of fully developed mass transfer can readily be formulated using fully developed heat transfer as a guide.

The Boussinesq transformation The combination of approximations and changes of variable that is utilized almost universally to simplify the equations of conservation for natural convection for small temperature differences, including, most importantly, the replacement of $-g - (\partial p/\partial x)/\rho$ by $g\beta(T - T_\infty)$, is generally called the Boussinesq transformation, although, according to David Hellums, he should be credited only in part. The approximations of the Boussinesq transformation can be, and increasingly are, avoided in numerical solutions, but at a significant cost computationally. The idealized solutions incorporating the Boussinesq transformation will probably continue to appear in our textbooks and even our handbooks because of their reasonable accuracy and the insight that they provide.

Asymptotic solutions for turbulent free convection Nusselt in 1915 speculated that the local heat transfer coefficient would eventually become independent of distance upward along a heated vertical plate. Such asymptotic behavior requires the proportionality of the local Nusselt number, Nu_x, to the one-third power of the local Grashof number, $Gr_x = g\rho^2\beta(T_w - T_\infty)x^3/\mu^2$. Fifty- five years later in 1970, I speculated in print that the inertial terms must become negligible relative to the viscous ones as $Pr \rightarrow \infty$ owing to the viscosity, and vice versa, implying proportionality to $Ra^{1/3}$ and $(RaPr)^{1/3}$, respectively. All three of these speculations appear to be validated by the somewhat limited experimental data. The numerically computed values of Shyy-Jong Lin of 1978, which are based on the κ-ε model, predict an overshoot (such as that of the orifice coefficient) before approaching the asymptotic proportionality of 1/3, and that probably explains the radically different powers obtained by various experimentalists.

The alternative speculation of Frank-Kamenetskii in 1937 of independence of the asymptotic heat transfer coefficient from both the viscosity and the thermal conductivity, which leads to a proportionality to $Gr^{1/2}Pr$, as well as other speculations that lead to a proportionality to $Gr^{1/3}$ rather than $Ra^{1/3}$, appear to be invalid. As an aside, Eckert and Jackson in 1951 derived, by means of integral-boundary-layer theory, an expression in which Nu_x is proportional to $Gr^{0.4}$. The exponent of 0.4 is simply an artifact of their arbitrary choice of a velocity distribution. One conclusion from this set of analyses is that all speculations need to be tested and a second conclusion is that later speculations are not necessarily better than earlier ones.

Ohm's laws Ohm in 1827 derived expressions for steady-state electrical conduction through resistances in parallel and in series. These expressions are regularly applied for combined thermal conduction and convection and also for surface-catalyzed chemical reactions controlled by mass transfer, adsorption, surface diffusion, and surface reaction in series and parallel (the so-called Hougen-and-Watson expressions).

The radiative heat transfer coefficient If thermal radiation occurs in series or parallel with thermal conduction, forced convection, and/or free convection, it is convenient to linearize the dependence on temperature by means of the following expression in order to allow the application of Ohm's laws, which are restricted to processes that are linear in the potential:

$$h_R \equiv \frac{\varepsilon\sigma(T_s^4 - T_s^4)}{T_s - T_\infty} = \varepsilon\sigma(T_s^4 + T_\infty^2)(T_s + T_\infty) \cong 4\varepsilon\sigma T_m^3$$

The exact value of T_m can be calculated for each specified set of values of T_s and T_∞, but, since the variation of the absolute value is constrained, some arbitrary fixed-mean value is usually employed as an approximation. A *radiative thermal conductivity* may also be defined but it is less useful.

Black-body and gray-body radiation The emission, absorption, reflection, and transmission of thermal radiation are a function of temperature as well as of wavelength for all fluids and solids. An exact treatment taking into account the spectral variations with wavelength is generally so complicated as to be impractical. However, the concepts of black and gray bodies have permitted the derivation of acceptable approximations in many instances, for example, by Bert Larkin in 1959 to characterize the contribution of radiation to heat transfer through insulations, and by John Chen in 1963 for packed beds and later for fluidized beds.

Special Forms of Transport

Fluidized beds As contrasted with the prior listed concepts and variables, those for fluidized beds have, because of its primary development in connection with continuous catalytic cracking, originated almost exclusively within chemical engineering. I will not attempt to summarize the developments in this topic, which continue to be produced to this day, but instead merely mention one early contribution that had a great personal impact. I had struggled to make sense of experimental data obtained from a pilot plant and then from the startup of the second full-scale fluidized-bed catalytic-cracker, and was overwhelmed by the elementary but perceptive analysis published a few years later in 1948 by Dick Wilhelm and Moosun Kwauk. For example, they noted that

1) the pressure drop at the point of fluidization is given by $(-\Delta P)_i = L_i(1-\varepsilon_i)g(\rho_s-\rho)$
2) the height of an expanded bed is given by $L(1-\varepsilon) = L_i(1-\varepsilon_i)$
3) the mean interstitial velocity can be expressed approximately as $u_{m0} = u_T\varepsilon^n$

Packed beds A review of the vast literature on flow through porous media reveals that the majority of the contributions have been by chemical engineers. I will represent this history by a single expression, namely that devised Sabri Ergun, a chemical engineer, in 1952:

$$\frac{-\Delta P}{L}\left(\frac{\varepsilon^3}{1-\varepsilon}\right)\frac{D_p}{\rho u_0^2} = 150\frac{\mu(1-\varepsilon)}{D_p u_9 \rho} + 1.75$$

Although the additive form and the two coefficients of the *Ergun equation* are empirical, the laminar and the inertial (not turbulent) terms both have a theoretical rational and have proven very successful as compound variables. As an aside, Noel deNevers used the concept of a hydraulic diameter to rationalize the laminar term rather successfully and the concept of a very rough pipe to rationalize the inertial term with less success, perhaps because the latter concept implied turbulent flow.

Laminar condensation Nusselt in 1916 derived the following solution in closed-form for the space-mean heat transfer coefficient for condensation in a laminar film by the ingenious choice of idealizations, namely an isothermal vertical plate of length L and temperature T_p, a pure saturated vapor at T_g, negligible heat capacity, inertia, and fully developed (parabolic) non-rippling flow for the condensate, and negligible drag for the vapor:

$$\Gamma\left(\frac{\mu_\ell \lambda^3}{g\rho_\ell^2 k_\ell^3 (T_g - T_p)^3 L^3}\right)^{1/4} = \frac{4^{3/4}}{3} = 0.9428$$

Here, Γ is the mass rate of condensation per unit breadth of the plate and λ is the latent heat per unit mass.

This solution can be interpreted as an ultimate example of the unique utility of a compound variable. Those of you who are scornful of such a purely theoretical result should ask yourself how much experimentation with what degree of accuracy would be necessary to produce such a relationship, that is lead to the identification of the power-dependence of the rate of condensation on each of these variables. Dimensional analysis would get you part way but not to this degree of resolution. Theoretical and/or numerical solutions have since been derived that avoid each of the idealizations of Nusselt, and they should be used for design. Even so, his solution remains a touchstone both because of the insight provided by its combination of variables and because of the fairly accurate quantitative agreement of its predictions with experimental data.

Some textbooks present this solution rearranged in terms of the Nusselt number (and thereby the heat transfer coefficient) as a function of the Reynolds number, but that is a misdirected effort because the rate of flow (the rate of condensation), which is a dependent rather than an independent variable, is needed to evaluate the Reynolds number.

Several Forms of Transport about Which You May Not Know

Thermoacoustic convection Incorporation of Fourier's law of conduction in the unsteady-state one-dimensional differential energy balance results in

$$\rho c \frac{\partial T}{\partial t} = \frac{\partial}{\partial x}\left(k \frac{\partial T}{\partial x} \right)$$

It has long been recognized by mathematicians that this model is faulty in that it predicts an infinite rate of propagation of energy for some thermal boundary conditions. Cattaneo in 1948, Morse and Feshbach in 1953, and Vernotte in 1958 independently proposed the following so-called *hyperbolic equation of conduction* to avoid that defect:

$$\rho c \frac{\partial T}{\partial t} + \frac{k}{u_T^2}\frac{\partial^2 T}{\partial t^2} = \frac{\partial}{\partial x}\left(k \frac{\partial T}{\partial x} \right)$$

Here, u_T is a the velocity of the thermal wave, usually taken to be the velocity of sound. This expression, which they rationalize on the basis of a misreading of the Maxwell's classical paper on the kinetic theory of gases, is pure rubbish with no theoretical or experimental basis, but I have identified over 200 solutions of it in the literature of applied physics and engineering, including a few by chemical engineers whom I will generously grant anonymity. It may be noted that liquids and solids, as well as gases, are sufficiently compressible to result in the generation of thermoacoustic waves.

Meanwhile, back in the realm of rationality – beginning in 1899 with Rayleigh, who postulating the waves to have a sinusoidal shape, and extending to the present, with numerical methods – thermal waves have been predicted with no heuristics by solving the differential energy, momentum, and mass balances as coupled by a variable density. Matthew A. Brown, with my urging, measured the velocity and shape of a wave generated by the sudden and extreme heating of the metallic end-surface of a gas-filled tube. The wave-shape disagreed decisively with all prior predictions, including our own. Upon close examination we detected some evidence of instability in the numerical calculations. It took a reduction of 10^4 in the steps in both space and time (10^8 in all) to eradicate that instability, but we were rewarded with agreement. The applications of thermoacoustic waves, other than thunder as generated by lightning) are mostly rather esoteric but this experience is a valuable reminder of the error and dangerous conclusions that may result from heuristic models, a lack of experimental measurements, misplaced faith in classical models, the misinterpretation of prior work, and/or insufficient tests of numerical computations.

The thermal conductivity of dispersions Maxwell in 1873, using the principle of *invariant imbedding*, derived a solution for the electrical conductivity of dispersions of spheres based on the postulate that the spheres are far enough apart so that deviations in the electric field generated by one sphere do not interfere significantly with those generated by adjacent ones. In 1986 I discovered that his solution, when re-expressed in thermal terms and rearranged in terms of only one compound dependent and one compound independent variable, represented, with minimal parametric dependence, the values determined from the closed-form solution of Rayleigh in 1892 and the numerical solutions of Howard Brenner in 1977, of Andy Acrivos in 1982, and others, for several ordered arrays. Even more remarkably, this solution provides a lower bound and a fair representation even for the extreme of a packed bed and even for granular materials. The corresponding solution for an array of long cylinders with their axes perpendicular to the direction of conduction, such as with fiberglass insulations, is readily formulated. This is a prime example of the utility of an identification of compound variables by means of a theoretical solution for a highly idealized condition.

The migration of water in porous media In order to test predictions of the rate of freezing of the soil outside underground tanks for the storage of liquefied natural gas (*LNG*) I persuaded Jai P. Gupta to measure, on the laboratory scale, the transient temperature produced by a flat, cold, bounding surface in sand with a representative concentration of water. He decided on his own to measure the transient concentration as well. As he reported in 1971, the experimental rate of heat transfer through the soil was in good agreement with theoretical predictions but the rate of moisture migration and thereby the rate of freezing was an order of magnitude greater than expected. My colleague David Graves conjectured, and we thereupon confirmed, that this rapid migration occurred in the liquid phase, was a consequence of the variation in surface tension of the water with temperature, and required the presence of a vapor phase. Consequent experiments with heating of the bounding surface produced an even greater surprise, and one with practical consequences in industrial drying, namely the migration of the water in the pore-space toward rather than away from a heated surface. We soon recognized that this reversal in the direction of the migration was a consequence of a maximum in the value of the surface tension of water at 4°C.

What is the description of this experience, which does not appear to involve either modeling or correlation, doing here? The answer is that it was the agreement of the measurements with the theoretical model for diffusion on the one hand and their failure on the other that called attention to their significance. The failure of the experimental results for heating to follow those for cooling had a similar consequence. In this instance, the combination of experimentation and theory produced two discoveries that neither would have done alone.

Similarity Transformations

A major factor in improved understanding and predictions in transport in laminar flow or the absence of flow has been the identification of similarity transformations, that is of combinations of variables that allow reduction of the order of the differential equation(s) of conservation, for example from partial to ordinary. In the dim past most of them were identified one by one and were considered to be signal achievements if not evidence of a stroke of genius. A sampling of familiar ones follows.

Transient thermal conduction Formal substitution of the variable $\xi = kt/\rho cx^2$ for t and x reduces the partial differential equation for transient thermal conduction in a semi-infinite region of an incompressible material with a uniform initial temperature and an imposed surface temperature to an ordinary differential equation that can readily be integrated. The origin of this transformation is unknown to me.

The thin boundary-layer transformation Prandtl in 1904 discovered a similarity transformation that reduced the partial-differential description of developing flow in the "thin laminar boundary regime" on a flat plate to an ordinary differential equation. This transformation led to series and numerical solutions for both flow and convection.

The Pohlhausen transformation E. Pohlhausen in 1921 discovered a similarity transformation that reduced the partial differential model for free convection on a vertical isothermally heated plate, in the thin-laminar-boundary-layer regime, to an ordinary differential model. It led to nearly exact numerical solutions thereof.

The Lévêque transformation Lévêque in 1928 ingeniously approximated the partial-differential model for laminar forced convection in the inlet region of a channel and discovered a similarity transformation that reduced that idealized model to an integral that could readily be solved numerically. The *Lévêque solution* is an invaluable asymptote for the regime near the entrance of a tube or channel, in which the Graetz solution converges too slowly to be useful and in which a finite-difference solution requires excessive discretization.

The Hellums-Churchill methodology In 1964, David Hellums, together with the speaker. devised a simple methodology that can be used to determine whether or not a similarity transformation is possible, and if so to derive it, thereby eliminating the need to be a genius such as Prandtl, Pohlhausen, and Lévêque. For example, the four aforementioned illustrative similarity transformations are each readily derived by undergraduates after only minimal instruction. Charles W. White, III, a graduate student of Warren Seider, on my urging, wrote a computer program that not only searches for similarity transformations but also performs the otherwise tedious reduction to an ordinary differential equation.

The integral transformation of Saville At my behest, Dudley Saville attempted to carry out a finite-difference solution for laminar free-convection from an unbounded horizontal cylinder of infinite length. He soon discovered that I had chosen an ill-posed problem – the buoyant flow generated by heating or cooling such a hypothetical object sets the whole universe in motion. However, this failure led to a great leap forward. He retreated to the thin-boundary-layer model and discovered a modification of the integral transformation of Görtler that led in 1965 to a common solution for cylinders of general contour in the form of an infinite series that converges more rapidly than the classical solution of Blasius for a round cylinder. He then utilized this methodology to develop the corresponding generalized solutions for horizontal cylinders and vertical axially symmetric bodies of general contour, and then for simultaneous heat and mass transfer.

Conventional Correlating Equations

Although theoretical concepts have had an ever-increasing role in the Century of the AIChE, the design of chemical plants and the analysis of chemical processing has from the beginnings to the present required the use of graphical representations or empirical correlating equations for both physical-chemical properties and for the processes themselves.

Dimensional analysis The correlating equations for transport have generally consisted of the product of powers of dimensionless groupings of dimensional variables. These dimensionless groupings are ordinarily determined by means of applying dimensional analysis to a listing of a sufficient set of variables, including physical properties, to characterize the process. It might appear unnecessary to review this topic, which is presumably well-known to all chemical engineers, but my experience indicates otherwise.

Rayleigh in 1915 got dimensional analysis, as applied to a listing of variables, exactly right. It is my personal opinion that the world would be better off if we simply ignored all subsequent contributions (except possibly as sources of examples). I can't take time today to describe his process in detail and will only note that he postulated an expression in the form of a power series of a term made up of the product of powers of all of the variables. He then focused his attention on only one of these terms and determined these powers insofar as they are constrained by the conservation of dimensions, including time. A typical solution might be

$$Nu = ARe^{n}Pr^{m} + BRe^{2n}Pr^{2m} + \ldots$$

Rayleigh realized that this result does **not** mean that the Nusselt number is proportional to the product of a power of the Reynolds number and a power of the Prandtl number. Rather it merely indicates that *Nu* is some unknown arbitrary function of *Re* and *Pr*, that is that

$$Nu = \varphi\{Re, Pr\}$$

Unfortunately, many chemical engineers, including academics, fail to make this distinction and then compound the error by plotting experimental data for *Nu vs Re* and/or *Pr* on log-log coordinates to determine values of *A*, *n*, and *m*. I wasted a significant amount of time in my younger years trying to rationalize the prevalent value of $n = 0.8$ as in the former expression because I confused it with 4/5. I eventually realized that meaningful powers of dimensionless groups only occur in asymptotes and that values such as 0.8 are rounded-off artifacts of the choice of some particular range of the variable (here *Re*) and have no theoretical significance.

If I convinced all or even some of you to abandon correlating equations in the form of arbitrary power functions and their products, as well as the use of log-log plots for their numerical evaluation, this would truly be a Centennial Day for you and for me. I have spent much of the last third of my professional life trying to make that choice possible, and plan to focus some of my remaining few minutes here to try to recruit you to help.

A Correlating Equation for Almost Everything

I am certain the recognition that power-dependences and their products occur only in asymptotes came to many of you before it did to me. However, it came to me as such an epiphany that I vigorously pursued the consequences; one of which was the further realization that exact or nearly asymptotes could be derived for most behavior, not only in transport, and not only in physics, but also in non-technical venues such as athletics and economics. It struck me that incorporating asymptotes in correlating equations might improve their accuracy both numerically and functionally. That proved to be so beyond all expectations. After some trial and error we arrived in 1972 at

$$y_0\{x\} = \left[y_0\{x\}^p + y_\infty\{x\}^p\right]^{1/p}$$

I eventually called this expression "the correlating equation for almost everything" in recognition of its universality, but have also referred to it as the *CUE* (for Churchill-Usagi equation) in the interests of brevity and in acknowledgement of my collaborator, Reneé Usagi.

The *CUE* can be seen to be the *p*th-power-mean of the two limiting asymptotes, $y_0\{x\}$ and $y_\infty\{x\}$. In order to evaluate the combining exponent, *p*, it is convenient to express the *CUE* in one or the other or both of the following canonical forms;

$$\frac{y\{x\}}{y_0\{x\}} = \left[1 + \left(\frac{y_\infty\{x\}}{y_0\{x\}}\right)^p\right]^{1/p} \quad \text{and} \quad \frac{y\{x\}}{y_\infty(x)} = \left[1 + \left(\frac{y_0\{x\}}{y_\infty\{x\}}\right)^p\right]^{1/p}$$

This expression is successful almost beyond belief because it reduces the task of correlation to that of representing the deviation from the closest asymptote rather than the whole variance. I will subsequently mention some examples that illustrate and confirm that assertion.

We were not the first to utilize this expression; earlier users include Andy Acrivos and Tom Hanratty. Our contribution was to recognize its potential for general usage and to devise a standardized procedure for determining the best value of the arbitrary exponent *p*. Statistical evaluation of the this exponent is unnecessary due to the insensitivity of the predictions to that quantity, and indeed a ratio of two integers may be chosen for convenience

In general the optimal value of *p* is arbitrary and without theoretical significance, but Eli Ruckenstein derived a theoretical value of 3 for combined free and forced convection, which, happily, was in agreement with the value we had already determined painstakingly from experimental data and numerically computed values.

As an aside, the aforementioned *Ergun equation* can be interpreted as an application of the *CUE* with *p* = 1. Several other examples are mentioned immediately below, and others subsequently in the treatments of analogies, turbulent flow, and turbulent convection.

Restrictions on the use of the *CUE*

Asymptotes must be known, derived, or formulated
Asymptotes must intersect once and only once
Asymptotes must both be upper or lower bounds
Asymptotes must be free of singularities

The behavior for large and small values must be symmetrical

Separate expressions are necessary for derivatives and for integrals, but derivatives and integrals of the asymptotes can be utilized

Examples

Laminar free convection The first utilization of the *CUE* in 1972 consisted of the following combination of the asymptotes of LeFevre of 1957 for the local Nusselt number in free convection from an isothermal vertical plate in the thin laminar boundary layer regime:

$$Nu_x^p = \left[0.6004\, Gr_x \mathrm{Pr}^{1/2}\right]^p + \left[0.5027\left(Gr_x \mathrm{Pr}\right)^{1/4}\right]^p$$

The right-most and middle terms are the asymptotic solutions for $Pr \to \infty$ and 0, respectively. A combining exponent *p* of 9/4 was found to represent experimental data and numerically computed values of Nu_x for a wide range of values of *Pr* and Gr_x within 1%.

With this combining exponent, the correlative expression can be reduced to

$$Nu_x = 0.5027\,(Gr_x Pr)^{1/4}/[1+(0.492/Pr)^{9/16}]^{4/9}$$

The denominator of this expression has proven to be such a good approximation for all geometries and thermal boundary conditions, and for the turbulent as well as the thin-laminar-boundary-layer regime, that it has been called "the universal dependence on the Prandtl number". That is excessive praise; slightly improved representations can be obtained with slightly different coefficients and combining exponents for each condition.

It should be noted that although the goal was a correlating equation for the dependence on the Prandtl number, the dependence on Gr_x was incorporated by virtue of its presence in the asymptotes. Furthermore, insertion of this expression for Nu_f in the previously presented equation of Langmuir provides a fair prediction for Nu_x for values of Gr_x below the range of thin-boundary-layer theory.

The orifice coefficient As an example of the application of the *CUE* of more general interest, we derived in 1974 an expression for the sharp-edged-orifice-coefficient for all values of *Re* using asymptotes for three regimes. For the asymptote for large *Re* we initially used 0.611. Empirical asymptotes for the regimes of creeping and transitional flow, as well as numerical values for the two combining exponents were initially derived from the 1933 experimental data of Tuve and Sprenkle for a_0/a = 0.4. The resulting predictive equation was subsequently improved by replacing the limiting value of 0.611 by the aforementioned exact value of 0.5793, and the coefficient of 0.16 by the exact value of $1/(12\pi)^{1/2}$ = 0.1629 from the solution of Roscoe in 1951 for creeping flow in the limit of $a_0/a \to 0$, which I belatedly learned about from Al Baer. Presumably, in the intervening years since 1933, the experimental measurements of Tuve and Sprenkle have been improved upon and extended to other values of a_0/a, or have been supplemented or replaced by numerical solutions, but I leave the up-dating to one of you, who may thereby place your name in future handbooks and textbooks.

Multiple variables Multiple independent variables may be taken into account with the *CUE* by either or both of two means. First, secondary variables may be incorporated in either or both of the asymptotes, as was Gr_x in the expression for laminar free convection, or by serial application. If expressions are known from theoretical considerations or can be contrived for consecutive regimes such as creeping, laminar, transitional, and turbulent flow, the application is usually straightforward except for the arbitrary choice of the order of combination. However, in many instances, the transition is sigmoidal in form and unknown. If so, it may be possible to represent the intermediate asymptote by an arbitrary power-function, which may or may not be the tangent through the point of inflection. Our first derivation of such representation – namely for the variation of the effective viscosity of a pseudoplastic fluid with the shear stress – was not only remarkably successful in a predictive sense but also suggested that the power-dependence of the intermediate asymptote (the Ostwald-deWaele model) may be a mathematical artifact rather than a physical phenomenon.

A generalized expression for transitional behavior In 1974, H.J. Hickman, a graduate student at the University of Minnesota, derived, using the *Laplace transform*, an infinite-series solution for fully developed convection in fully developed laminar flow with heat losses due to radial conduction through the tube wall and external insulation, in series with free convection and thermal radiation to the surroundings. In 2001, I recognized that this solution, after the correction of small numerical errors, could be represented exactly by a highly reduced form of the *CUE*, namely

$$\frac{Nu_J - Nu_T}{Nu \ - Nu_T} = 1 + \frac{Bi}{Nu_T}$$

Here Nu_J *and* Nu_T are the internal Nusselt numbers for uniform heating and a uniform wall-temperature, respectively, $Bi=U_eD/k$ is the Biot number, and U_e is the overall heat transfer coefficient for the heat losses, I further recognized that an expression of this general form could be used to represent many other transitional processes, an example of which is presented subsequently in connection with turbulent convection in a tube.

Status and future of the *CUE*

It goes without saying that graphical correlations are more or less incompatible with computer-aided design, although they remain important to reveal the quality of the representation.

A collateral observation is that most of the scatter observed in conventional log-log plots is due to mis-correlation not poor data.

The *CUE* appears to provide useful replacements for those graphs and power-law representations.

In the past 36 years, my students and other collaborators have developed several hundred predictive or correlative expressions in this form, primarily for transport. I am gratified that several recent books on heat transfer present, almost exclusively, the expressions that we have devised, but am disappointed that so few investigators have followed our lead in developing expressions in this form for their own results. I hope that some of you will be inspired to do so.

Analogies

Many of the analogies in common use by chemical engineers were first conceived by scientists and other brands of engineering. Even so, analogies appear to have a more essential role in chemical engineering, perhaps because of the breadth of our interests and the consequent involvement with processes in which flow, heat transfer, mass transfer, and chemical reactions occur simultaneously. The differential equations of conservation for momentum, energy, and species display some commonalities that suggest analogous behavior but they also display fundamental differences. Most of the analogies now in common use sprung from some ingenious insight rather than by comparison of the equations of conservation. I will start with simple ones, go on to more complex ones, and then describe in some detail the development of a relatively new one.

The equivalent diameter Perhaps the most widely used analogy in practice is that of the equivalent diameter. This concept allows correlations for flow and/or heat transfer in one geometry, usually a round pipe, to be utilized as an approximation for other geometries. The concept requires the choice of some arbitrary methodology to determine the equivalent diameter. The most common choice is the *hydraulic diameter*, which is defined as four times the cross-sectional area divided by the wetted-perimeter. The *hydraulic diameter* results in an over-prediction of the friction factor for laminar flow in a parallel-plate channel by 50% but a lesser error in turbulent flow and in annuli and rectangular ducts. The *laminar-equivalent* diameter, that is the value that produces the exact value for laminar flow, is somewhat more accurate but requires the availability of a solution or correlation for the latter.

The analogy of MacLeod Bob Rothfus and Carl Monrad determined the conditions that force the velocity distribution in fully developed laminar flow in round tubes and parallel-plate channels to be congruent. Remarkably these conditions are exactly equivalent to expression of the results for round tubes in terms of and a^+ and y^+, and those for parallel-plate channels in terms of y^+ and b^+. This result is intriguing but of little practical importance. They credit a doctoral student, Alexander MacLeod, with the conjecture that this congruence might carry over to turbulent flow. A plot of experimental data by Glen Whan and Bob Rothfus confirmed that conjecture and also demonstrated that it does not apply to the regime of transition. This discovery of congruence is of great importance not only because it allows experimental data for the two geometries to be used interchangeably but even more importantly because it allows the numerical values obtained for parallel-plate channels by *DNS* to be applied for round tubes.

The analogy between heat and mass transfer The analogy between heat and mass transfer is based on the similarity of the equations of conservation. Insofar as they are identical in form, the Sherwood number and the Schmidt numbers can simply be substituted for the Nusselt number and the Prandtl number, respectively, in a correlation or solution for heat transfer. That is useful because the correlations and solutions for heat transfer are much more extensive. However, the similarity may be spoiled by a number of factors, principally the bulk motion generated by non-equimolar mass transfer, the dependence of the Schmidt number on composition, and the differing dependences of the Schmidt and Prandtl numbers on temperature and pressure. It may be noted as an aside that in 1997, Dimitrios Papavassiliou and Tom Hanratty explained, using *DNS*, the failure of the analogy for large values of *Sc*.

The analogy between electrical and thermal conduction The analogy between electrical and thermal conduction is exact for steady state processes but doesn't apply to unsteady state ones. It has few active applications but two are of great importance. Maxwell's law of dispersions and Ohm's law were conceived in electrical terms but their analogues in chemical and thermal processing have proven to be of great utility.

An analogy for buoyant processes Emmons in 1954, recognizing that laminar free convection, film condensation, film boiling, and film melting are all controlled by thermal conduction across a film moving at a velocity controlled by gravity, derived the following generalized solution:

$$\frac{h_m L}{k} = \left(\frac{\rho F L^3}{K \mu k \alpha \Delta T} \right)^{1/4}$$

Here, F is the gravitational force per unit volume of the film, K is a numerical factor characterizing the viscous shear, and α is the increase in the mass rate of flow of the film per unit of heat transfer. He tabulated values of the factors F, K, and α for each of the four processes on the basis of qualitative considerations. One merit of this analogy is that it allows a common plot of experimental data for all four processes. More accurate coefficients have been derived for each of the four individual processes so the inaccuracy of the tabulated values of Emmons represents an intentional sacrifice to obtain commonality and insight. As an aside, the term in brackets, after substitution for F, K, and α, may be identified as the Rayleigh number in the case of free convection but the counterparts for the other three gravitational processes do not appear to have an accepted "name".

The corresponding analogy for the turbulent regime is

$$\frac{h_m L}{k} = A\left(\frac{\rho F L^3}{K \mu k \alpha \Delta T}\right)^{1/3}$$

Here, A is an empirical coefficient. Emmons' values of F, K, and α convert this expression to the accepted form for turbulent free convection, but the limited and very scattered experimental data for film condensation, film boiling, and film melting in the turbulent regime do not appear to confirm the predicted independence from L.

In a qualitative if not a quantitative sense, the analogy of Emmons represents one of the broadest generalities in all of transport.

An analogy between chemical reaction and convection Although energetic chemical reactions have been known for over 40 years to enhance or attenuate convection radically this important aspect of behavior does not appear to have found its way into any textbooks or handbooks on heat transfer or mass transfer. The earliest studies are by well-known chemical engineers including P.L.T. Brian, Bob Reid, and Samuel Bodman in the period 1961-1965, Joe Smith in 1966, and Louis Edwards and Robert Furgason in 1968. My encounter with this phenomenon in combustion beginning in 1972, and the absence of a simple predictive expression, eventually led me to derive one. First, I derived for a very idealized set of conditions

$$Nu = Nu_o/(1+\beta Q)$$

Here Nu_o is for no reaction, Q is the ratio of the heat of reaction to the heat flux from the wall, and β is an arbitrary coefficient. This expression led by a path too long and complex to describe here to the following one relating the local mixed-mean rate of reaction, as represented by $\xi(1-Z_{mx})$, to the local rate of heat transfer from the wall, as represented by Nu_x:

$$Nu_x = \frac{Nu_{ox}}{1+\beta\xi(1-Z_{mx})\exp\left\{\dfrac{E/RT_0}{1+\left(\tau\left[Z_{mx}+\left(K_{0x}/\xi\right)\right]\right)^{-1}}\right\}}$$

This expression, which is restricted to fully developed laminar or turbulent flow with a uniform heat flux on the wall and a first-order reaction, explains as well as predicts almost exactly the the extreme and chaotic enhancement and suppression of the Nusselt number. It differs fundamentally from conventional analogies in that it is for a developing rather than a fully developed process.

Turbulent Flow

I have three reasons for giving this topic more attention than any other. First, most chemical processing takes place in turbulent flow, second, turbulent flow and convection best illustrate the advance of our technical capability over the past 100+ years, third, chemical engineers have had a critical role in its evolution, and fourth, the state of the art in dealing with turbulent flow may not be known to all of you. Most of the topics that follow qualify for the prior section on concepts or that on analogies. They were deferred to this section in the interests of continuity.

Time-averaging In my view, the greatest advance of all time in modeling turbulent flow and transport in turbulent flow was by Osborne Reynolds, who in 1895 space-averaged the partial differential equations of conservation. The equivalent of this space-averaging is now carried out by time-averaging. Not everyone agrees with my assessment because the process produces some new effective variables such as $\rho\overline{u'v'}$ and $\rho c\overline{v'T'}$, for which arbitrary predictive expressions are required. Even so, all turbulent modeling for nearly a century started with these time-averaged expressions and they may still have some applicability in the future.

The eddy diffusivity In 1877, (and thus before the turbulent shear stress, heat flux density, and mass flux density were defined in terms of the fluctuating components of the velocity, temperature, and composition by Reynolds) Boussinesq introduced the eddy viscosity, eddy conductivity, and the eddy diffusivity to model them. They could have been examined in the section on analogies because Boussinesq conceived of these expressions for turbulent transport as analogs of those for molecular diffusion. I will try to summarize as briefly as possible the path that led us from Reynolds and Boussinesq to where we are today. That is not easy because it involves as many missteps and side-steps as advances.

The power-law for the velocity distribution In 1913, Blasius, a student of Prandtl, fitted a log-log plot of the available data for the friction factor versus the Reynolds number in a round pipe with a straight line of slope of –1/4. Prandtl presumed this nice round fraction had some significance, and in 1921 derived from it, with extraordinary ingenuity, a one-seventh-power dependence for the velocity on the radius. He subsequently realized, in part due to the obvious deficiencies of the one-seventh power velocity distribution, that the slope –1/4 was merely an artifact of the narrow range of the experimental data available to Blasius.

The mixing-length Having found the power-law to be deficient, Prandtl in 1925 next proposed to represent the turbulent shear stress with $\rho\ell^2(du/dy)^2$, where ℓ is a *mixing-length* for the turbulent eddies analogous to the mean-free-path of gaseous molecules. Bird, Stewart, and Lightfoot, many years later in 1960, but while this concept was still in vogue, correctly criticized it as a "very poor analogy". Prandtl conjectured that the mixing length might vary linearly with distance from the wall near the wall, that is, $\ell = ky$, and used this conjecture, along with several clever idealizations to derive

$$u^+ = A + \frac{1}{k}\ln\{y^+\}$$

Because of its obvious failure at the wall and at the centerline, the region of applicability of this expression has become known as the "turbulent core near the wall". Because of his subsequent, presumably independent derivation of the same expression, the empirical coefficient of linearity, *k*, is generally known today as the *von Kármán constant*. Von Kármán and Prandtl both further conjectured that, in spite of its failures near the wall and near the centerline, the integration of this expression over the cross-section might yield a good approximation for the mixed-mean velocity and thereby the friction factor, namely

$$u_m^+ = \frac{u_m}{\sqrt{\rho\tau_w}} = \sqrt{\frac{2}{f}} = A - \frac{3}{2k} + \frac{1}{k}\ln\{a^+\} = B + \frac{1}{k}\ln\{a^+\}$$

This derivation is one of the most fateful in history because the resulting expression has by means of a variety of empirical values of *A* (or *B*) and *k*, remained in our handbooks, either algebraically or graphically, usually without the inclusion of supporting data, to this day.

Prandtl in 1925 further conjectured that the mixing length might approach a constant value at the centerline leading to the equivalent of the following erroneous asymptote for that region:

$$u = u_c - Cr^{3/2}$$

In order to encompass a wider range of behavior, von Kármán in 1930 proposed the following theotetical-looking expression:

$$\ell = k\left(\frac{du/dy}{d^2u/dy^2}\right)$$

I once asked him, over a glass of champagne, the source of this, to me, mysterious relationship. Without batting an eye, he replied that it was the simplest expression, involving only derivatives of the velocity, with the correct overall dimension. After considerable analysis, I satisfied myself that this was true. Unfortunately, however, the predictions of this elegant expression are quite erroneous and misleading because of the failure of the very concept of a mixing length.

Wall-based variables In 1926, the year after proposing the mixing length, Prandtl took another tack. First he used routine dimensional analysis to derive

$$u\sqrt{\rho/\tau_w} = \varphi\{y\sqrt{\rho\tau_w}/\mu, a\sqrt{\rho\tau_w}/\mu\}$$

Then he introduced the following symbols which have remained in use for over 80 years

$$u^+ = \varphi\{y^+, a^+\}$$

The law of the wall Prandtl next conjectured that near the wall the dependence on a^+ would phase out leading to

$$u^+ = \varphi\{y^+\}$$

which is known as "the universal law of the wall". This is a seemingly straightforward application of dimensional and asymptotic analysis but it is not. Had the pressure gradient rather than the shear stress on the wall, or the radius rather than the distance from the wall been chosen as variables, or had some other grouping such as $ua\rho/\mu$ been chosen as the dependent variable, the elimination of *a* would not have led to such a useful result. Although he did not say so in print, my guess is that Prandtl tested and then rejected each of these alternatives as inferior.

The law of the centerline Postulating that the velocity field near the centerline to be independent of the viscosity next lead Prandtl to the following counterpart for this region:

$$u_c^+ - u^+ = \varphi\{a/y\}$$

The term $u_c^+ - u^+$ is called the *velocity defect.*

The law of the turbulent core Millikan in 1938 ingeniously recognized that the only expression that conformed to both of these two functional asymptotes was

$$u^+ = B + \frac{1}{k}\ln\{y^+\}$$

This derivation, which is free of any heurism, reveals that two erroneous concepts (the mixing length and its linear variation near the wall) may lead to a valid result.

The limiting behavior at the wall Prandtl also postulated that very, very near the wall the shear stress due to the turbulent fluctuations and the effect of curvature would be negligible leading to

$$u^+ = y^+$$

This expression can be noted to conform to "the law of the wall."

The κ-ε model The best-known predictive equation for the eddy viscosity is the κ-ε model of Kolmogorov, Prandtl, and Batchelor. The *kinetic energy of turbulence* κ and the *rate of dissipation of turbulence* ε have physical significance but the equations devised beginning in 1972 by Launder and Spalding and others for their prediction are very empirical, based on homogeneous turbulence, and not very accurate functionally or numerically, particularly near the wall where the predictions are most important. This model has been superseded for fully developed flow but remain the principal, if unreliable, source for developing flow.

Direct numerical simulation (*DNS*) Prior to its accomplishment, I conjectured that numerical integration of the time-dependent equations of conservation for the turbulent regime was not feasible because the scale of the turbulent eddies extends from that of the size of the channel down to that of the molecules, and that the grid-size would need to extend down to the latter level. My reasoning was faulty because I failed to realize that the grid-size only needs to be decreased to the point at which the turbulent shear stress is negligible compared to the viscous shear stress. In an informal conversation at the AIChE Annual meeting in 1990 I learned from Charlie Sleicher and Tom Hanratty that they and their doctoral students had followed the lead of Kim, Moin and Moser in 1987 and carried out such calculations, and that they were essentially free from empiricism. I immediately jumped to an even more erroneous conclusion, namely that we had entered a new era in which such calculations would replace all other approaches to turbulent flow and convection. However, after more than 20 years, *DNS* calculations are essentially limited to planar flows at rates barely above that for fully developed turbulence.

Large-eddy simulation (LES) This methodology, as devised by Schumann in 1975, relaxes the restriction on the rate of flow by utilizing *DNS* only for the fully turbulent core, but requires the use of the κ-ε model with arbitrary wall-functions or the equivalent for the region near the wall.

We sorely need a new algorithm or concept that will deliver the predictions of turbulent flow promised but not produced by *DNS* and *LES*.

The turbulent shear stress very near the wall In 1932, Eger Murphree, a chemist, and somewhat later, Charlie Wilkie, a chemical engineer, as well as others, proposed on the basis of speculative asymptotic analysis of the time-averaged equation for the conservation of momentum that very near the wall the expression

$$-\rho\overline{u'v'} = \alpha(y^+)^3 + \beta(y^+)^4 + \cdots$$

The existence or non-existence of the term in $(y^+)^3$ was disputed for over 50 years because of the difficulty of carrying out experimental measurements of the required accuracy. This third-power dependence is of critical importance because it implies a proportionality of Nu to $Pr^{1/3}$ rather than $Pr^{1/4}$ in the limit of $Pr \rightarrow \infty$. The issue was finally settled definitively in favor of $(y^+)^3$ by the earliest results of *direct numerical simulation* (*DNS*), including those of Rutledge and Sleicher and of Lyons, Hanratty, and McLaughlin. This result implies that the velocity distribution near the wall is represented by

$$u^+ = y^+ - \frac{\alpha(y^+)^4}{4} + \cdots$$

The afore-mentioned calculations also provided the first reasonably accurate value for the coefficient α, namely 0.00070.

The local fraction of the shear stress due turbulence As an alternative to the representation of the turbulent shear stress by heuristics such as the eddy viscosity and the mixing length, Christina Chan and I in 1995 proposed to represent it directly – at first in terms of $(\overline{u'v'})^+ \equiv -\rho(\overline{u'v'})/\tau_w$, but later in terms of $(\overline{u'v'})^{++} \equiv -\rho(\overline{u'v'})/\tau$. The latter has the advantage of being finite at the centerline. From their definitions, $\dfrac{\mu_t}{\mu} = \dfrac{(\overline{u'v'})^{++}}{1-(\overline{u'v'})^{++}}$. This result confirms that, despite its heuristic origin and thereby the contempt of many "purists", the eddy viscosity really has physical significance, that is, the turbulent shear stress is proportional to the velocity gradient. (Boussinesq was either lucky or very insightful.) At the same time, the eddy viscosity is inferior in terms of simplicity and singularities to $(\overline{u'v'})^{++}$ and is therefore now of historical interest only. If you are still using it as a variable for fully developed flow you are out of date.

Expressing the mixing length in terms of $(\overline{u'v'})^{++}$ reveals that, despite all the criticism it too is independent of its mechanistic and heuristic origin. However, it is also revealed to be unbounded at the centerline of a round tube or the central plane of a parallel plate channel. How did such an anomaly escape attention for more than 70 years. One explanation is the uncritical acceptance by Prandtl of the plot of values of the mixing length obtained from the "adjusted" experimental values of Nikuradse, followed by the uncritical extension of respect for Prandtl and von Kármán to all of their derivations.

An algebraic correlating equation for the turbulent shear stress Eventually in 2000 we devised, using the *CUE*, the following expression for the local fraction of the total shear stress due to turbulence:

$$(\overline{u'v'})^{++} = \left(\left[0.7\left(\frac{y^+}{10}\right)^3\right]^{-8/7} + \left|\exp\left\{\frac{-1}{0.436y^+}\right\} - \frac{1}{0.436a^+}\left(1+\frac{6.95y^+}{a^+}\right)\right|^{-8/7}\right)^{-7/8}$$

The term inside the square brackets is the asymptote for $y^+ \to 0$, and that inside the absolute value signs that for $y^+ \to a^+$. An expression for the extreme limiting behavior of $(\overline{u'v'})^{++}$ for that condition was derived from the observed behavior, but was subsequently recognized to have the same form as one derived by Hinze in 1963 on the basis of the observation of a finite value of the eddy viscosity at the centerline. The numerical coefficients and the exponent of −8/7 are based on the experimental data of Zagarola. (Note the value of 0.436 for the *von Kármán constant.*) According to the aforementioned analogy of McLeod, this expression for $(\overline{u'v'})^{++}$ is applicable for parallel–plate channels if b^+ is substituted for a^+. We have also adapted it for circular concentric annuli.

The ultimate correlating and predictive equation for the friction factor Essentially exact expressions for the local and mixed-mean velocities follow directly from that for $(\overline{u'v'})^{++}$. First, the differential momentum balance is integrated to obtain values for $u^+\{y^+, a^+\}$ and $u_m^+\{a^+\}$ in terms of $(\overline{u'v'})^{++}$. These values are then used to evaluate the coefficients and combining exponents for correlating equations constructed from integrals of the asymptotic terms of the correlating equation for $(\overline{u'v'})^{++}$. The resulting expression for u_m^+ and thereby for the friction factor in a round tube is

$$u_m^+ = \sqrt{\frac{2}{f}} = 3.3 - \frac{227}{a^+} + \left(\frac{50}{a^+}\right)^2 + \frac{1}{0.436}\ln\left\{\frac{a^+}{1+0.301(e/a)a^+}\right\}$$

The resulting expressions for the local turbulent shear stress and the local time-mean velocity, as well as this one for the mixed-mean velocity (equivalent to the friction factor), represent the experimental data and the essentially exact numerically computed values for these quantities within their uncertainty. The unfamiliar terms in $227/a^+$ and $(50/a^+)^2$ arise from the boundary layer, which has been neglected explicitly in all prior integral expressions.

An iterative solution is required to determine the friction factor for a specified value of $Re = 2a^+u_m^+$ but the convergence is rapid, and the gain in accuracy, relative to expressions explicit in *Re*, more than justifies the necessity of the iteration. (This conclusion of mine is regularly contested, most recently in the March 2008 issue of *Chemical Engineering Progress*, but I am unconvinced that iteration is unacceptable or even onerous.)

The following expression in the form of two consecutive applications of the *CUE* predicts the friction factor for a round tube for all regimes of flow, including transitional, and all effective roughness ratios:

$$f = (f_l^{12} + [f_t^{-16} + f_T^{-16}]^{-3/2})^{1/12}$$

Here, $f_\ell = 16/Re$ (Poiseuille's law), $f_t = (Re/37530)^2$, and f_T is the above expression for fully turbulent flow. This expression is a complete replacement for and improvement on all expressions and plots for the friction factor. Although it obviates the need for one, it is can readily be programmed to produce such a plot in every detail.

Experimental Data

The most coherent and comprehensive set of experimental data in the history of transport is that of Nikuradse in 1930, 1932, and 1933 for the velocity distribution and pressure drop in flow through piping. He recognized that the failure of his measurements to conform to the theoretical prediction by Prandtl of $u^+ \rightarrow y^+$ for $y^+ \rightarrow 0$ must be due to experimental error. Unfortunately, rather than repeat his measurements with improved instrumentation he "adjusted" the values to fit the theory of his doctoral advisor. Had Nikuradse not done so, Prandtl would presumably recognized the fundamental failure of the concept of a mixing length 40 years before Christina Chan and I proved it to be fundamentally in error in even a mathematical sense.

Although many measurements of the velocity distribution and pressure gradient were made through the years, a comprehensive set of data to replace these "adjusted" values outright had to wait for Zagarola in 1996, and his were only for smooth pipe and did not extend quite to the lower limit of the turbulent regime. Despite heroic efforts by Zagarola and his colleagues in the Astronomy department at Princeton University to polish the surface, I conclude that even at the highest Reynolds numbers he attained,, the roughness of his tubing affected the measurements.

Turbulent Convection

As contrasted with turbulent flow, whose exiguity I have recounted in terms of asymptotics and speculations, the history of turbulent convection unfolds almost wholly through analogies between momentum and energy transfer. Literally dozens of such analogies have been proposed, most incorporating some novelty and/or minor advancement. I have chosen a limited, arbitrary set of those that have served as primary quantitative predictors of turbulent convection. However, before turning to the analogies, one solution and one correlating equation, and several quantities and relationships need to be identified.

The solution of Sleicher As most of you know, Graetz in 1883 devised an infinite series solution for developing onvection in fully developed laminar flow following a step in wall temperature. Despite its limitations in a practical sense, the Graetz solution is one of the icons of thermal science. Charles Sleicher in 1956 devised the equivalent series for turbulent flow and convection using empirical correlating equations for the velocity distribution and eddy diffusivities, and an analog computer to determine the eigenvalues and eigenfunctions. In 1969, Notter and Sleicher greatly improved that accuracy of that solution using a digital computer and updated expressions for the velocity distributions and eddy diffusivities.

Generalized correlating equations for forced convection In 1977, I devised, using the *CUE* with 5 asymptotes and four combining exponents, a correlating equation for *Nu* and *Sh* for all *Pr* and *Sc*, respectively, and all *Re* (including the laminar, transitional and turbulent regimes). The same structure, but with different coefficients, was proposed for uniform heating and uniform wall-temperature. A graphical comparison of these expressions with experimental data and a few numerically computed values of Sleicher and Notter revealed that, except for mass transfer for very large values of *Sc*, the discrepancies are generally within the uncertainty of the data. Those discrepancies are now presumed to be due to a breakdown of the analogy between heat and mass transfer. Complementary expressions were also devised for developing convection. The asymptotes for *Nu* and *f* for the turbulent regime have been superseded by the afore-presented improved expressions, but the overall structure and the overall correlating equation remain valid.

The eddy conductivity and the turbulent Prandtl number The turbulent conductivity ratio is analogous to the eddy viscosity ratio, and the fraction of the local heat flux density due to the turbulent eddies, namely, $(\overline{T'v'})^{++} \equiv \rho c_p(\overline{T'v'})/j$, to $(\overline{u'v'})^{++}$. It proves convenient to introduce still another analogous variable, namely the turbulent Prandtl number ratio,

$$\frac{\Pr_t}{\Pr} = \frac{(\overline{u'v'})^{++}[1-(\overline{T'v'})^{++}]}{(\overline{T'v'})^{++}[1-(\overline{u'v'})^{++}]}$$

in place of $(\overline{T'v'})^{++}$ in the differential energy balance. Although Pr_t varies with *Pr* and with $(\overline{u'v'})^{++}$, its variation is more constrained than that of $(\overline{T'v'})^{++}$.

Peter Abbrecht in 1956 determined the eddy conductivity experimentally in a developing temperature field and thereby confirmed his own conjecture that it was independent of the temperature field and thereby of the thermal boundary condition. It follows that Pr_t/Pr and $(\overline{T'v'})^{++}$ are as well. Furthermore, from the analogy of MacLeod it can be inferred that k_t/k, $(\overline{T'v'})^{++}$, and Pr_t /Pr, are identical for a round tube and a parallel plate channel when expressed in terms of a^+ and b^+, respectively. The limited experimental data appear to confirm this conjecture.

The Reynolds analogy Osborne Reynolds in 1874 postulated that momentum and energy were transported at equal mass rates from the bulk of the fluid to the wall by the oscillatory radial motion of turbulent eddies and thereby obtained a result that can be expressed in modern terms for a round tube as

$$Nu = PrRef/2$$

or in terms of individual dimensional variables as

$$h = 2\tau_w c/u_m$$

The latter expression reveals that the postulates of Reynolds imply independence from the diameter as well as from the viscosity, density, and thermal conductivity. The *Reynolds analogy* is noteworthy, not only for its date of appearance and its freedom from any specific mechanism of turbulent transport, but most importantly for its semi-quantitatively correct predictions. Remarkably, it remains implicit in the latest expressions for turbulent convection.

The Prandtl-Taylor analogy Prandtl in 1910 and G.I.Taylor, independently in 1916, attempted to correct for the major conceptual shortcoming of the Reynolds analogy by postulating that the eddies penetrate only to a finite distance from the wall, δ_s, called the *laminar sublayer thickness*, and that the transport of momentum and energy across that distance occurs by molecular diffusion. Their solutions were for unconfined flow along a flat plate but they can be rewritten approximately as follows for a round tube:

$$Nu = \frac{\Pr \operatorname{Re} f/2}{1+\delta_s^+(\Pr-1)(f/2)^{1/2}}$$

Max Jakob in 1949 remarked that "There exists a whole literature about (the equivalent of δ_S^+)." However, even a fixed numerical value of say 11 provides a reasonable prediction of *Nu* for all values of *Re*. Although the *Prandtl-Taylor* analogy has the merit of correctly predicting a shifting proportionality of *Nu* from *Ref/2* to $Re(f/2)^{1/2}$ as *Pr* increases, it shares the shortcomings of the Reynolds analogy for $Pr \leq 1$, and erroneously predicts independence from *Pr* for $Pr \rightarrow \infty$.

The Reichardt analogy Reichardt in 1951developed a greatly improved analogy by dividing the one-dimensional differential momentum balance in terms of the eddy viscosity ratio by the equivalent differential energy balance, thereby eliminating y^+, the dimensionless distance from the wall. He then introduced a potpourri of ingenious physical and algebraic approximations that allowed him to integrate the resulting expression and obtain a closed-form solution for *Nu* in terms of *Re, Pr,* and *f*. The *Reichardt* analogy, although far more accurate functionally and numerically than that of Prandtl and Taylor, has not received much direct usage because it is also far more complicated, requiring empirical graphical representations for its several variable coefficients. Its structure has however been utilized in most subsequent analogies, including those of Bill Friend and Art Metzner in 1958, Petukhov in 1970, Volker Gnielinski in 1976, and my own in 1997.

A reinterpretation and improvement of the Reichardt analogy Churchill, Shinoda, and Arai in 2000 noted that the Reichardt analogy could be interpreted as an interpolating equation in the form of the *CUE*, thereby greatly simplifying its expression and its implementation. Churchill and Zajic in 2001 took advantage of this reinterpretation to devise greatly improved analogies for all values of *Pr* and *Re*. As an example, their expression for $Pr \geq Pr_t$ is

$$\frac{1}{Nu} = \left(\frac{\mathrm{Pr}_t}{\mathrm{Pr}}\right)\frac{1}{Nu_1} + \left[1 - \left(\frac{\mathrm{Pr}_t}{\mathrm{Pr}}\right)^{2/3}\right]\frac{1}{Nu_\infty}$$

Here, $Nu_1 = \beta Ref/2$ and $Nu_\infty = 0.07343\left(\frac{\mathrm{Pr}}{\mathrm{Pr}_t}\right)^{1/3}\mathrm{Re}\left(\frac{f}{2}\right)^{1/2}$. The presence of the very first analogy (that of Reynolds) in this latest analogy should be noted. The coefficient β is an exactly defined function of the velocity distribution but the expressions representing this functionality, which differ for uniform heating and uniform wall-temperature, are too complicated to include here. The coefficient of Nu_∞, namely $3^{3/2}(0.0007)^{1/3}/2\pi = 0.07343$, is empirical only by virtue of its incorporation of the coefficient of 0.0007 in the previously given expression for $(\overline{u'v'})^{++}$ very near the wall.

The Colburn analogy The best-known analogy between momentum and energy transfer to chemical engineers is that derived by Alan Colburn in 1933. As contrasted with all of those I have just described, it has no mechanistic or theoretical basis. With great insight he observed a possible functional and numerical similarity between the following empirical correlating equations of E.C. Koo, a doctoral student at MIT, for the friction factor, and of Dittus and Boelter for the Nusselt number:

$$f = 0.046/Re^{0.2}$$

$$Nu = ARe^{0.8}Pr^n$$

The recommended values of *A* and *n* were 0.0243 and 0.4 for heating the fluid, and 0.0265 and 0.3 for cooling it. (As an aside, the gross functional inadequacy of the Dittus-Boelter equation was first exposed by its failure to predict heat transfer for liquid metals when they came into widespread use in connection with nuclear reactors in the 1950s.)

Colburn recognized that taking the ratio of these two equations would result in one in which *Nu* was proportional to *Ref*. In order to force the net coefficient to be an integer, he chose A = 0.023 for both heating and cooling, which was quite a stretch, and as a compromise between 04 and 0.3 chose n = 1/3. He later disclaimed any theoretical basis for that choice. The final result was

$$f/2 = Nu/RePr^{1/3}$$

He named the grouping on the right-hand side the *j-factor*, because of its analogy to the *f-factor* (*friction factor*), giving it a life of its own. This expression, together with an empirical correlating equation for the friction factor, remains in use to this day, although it is seriously wrong functionally in every respect, and, as I will now show you, numerically as well.

A graphical comparison of analogies for turbulent convection

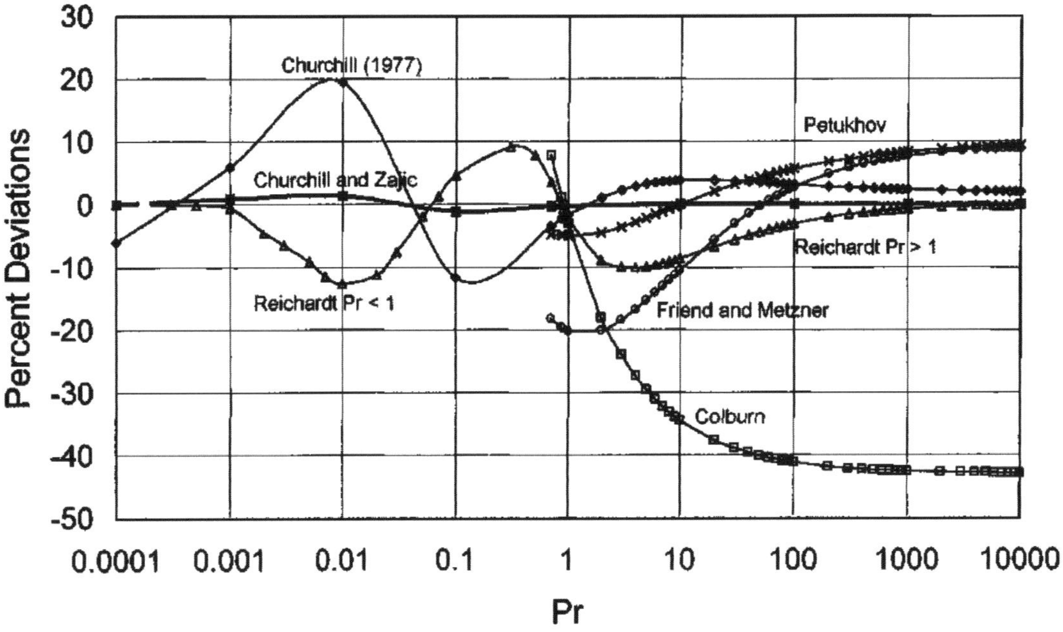

This plot of the percent error of the various predictive expressions for turbulent convection, as compared with essentially exact numerically computed values, reveals that the Colburn analogy is grossly in error except for *Pr* = 1, and that all of the analogies except that of Zajic and myself are in modest error for most values of *Pr*. Although this plot is for Re = 227,000 (*a*+= 5000), we obtained similar results were for all values of *Re* above 4000.

I have asked a number of experts in process heat transfer if they were aware that the predictions of the Colburn analogy were in such error. The most common answer is that they know the predictions of their computer package for heat transfer are grossly low, and that they compensate by applying a personal safety factor. I believe it is not disrespectful to Alan Colburn, whom I knew and greatly admired, to suggest the discard of a relationship that has out-lived its usefulness.

Devising Algorithms

Without stepping on the toes of Bruce Finlayson, I would like to mention some contributions of chemical engineers to numerical computations that I presume to be are outside the scope of his talk. Advances in computer hardware and software, including the development of computational algorithms for simulation, are not the whole story. Conceptual advances as primed by need have played a significant role in the advancement of transport. A few examples follow.

Electronic computational machinery first became available at the very time that I began my professorial career in 1952, and I subsequently have asked all of my doctoral students to undertake numerical solutions as well as closed-form solutions and experimental measurements. The state of the computational art was not only primitive when we began, but, despite all the formal advances, has to this day lagged behind their needs, forcing them to devise their own numerical algorithms. As an illustration of the ingenuity of chemical engineers in this respect, I will first mention a few such contributions primarily in just one narrow form of transport, namely natural convection in enclosures.

In 1952, when a "large-scale" electronic digital computer became available to us, William R. Martini asked if I believed that the partial differential equations of conservation described the real world and if we solved them numerically would we obtain accurate descriptions thereof. With some bravado, I replied "Certainly", leading him to attempt the first-ever two-dimensional computations for confined natural convection. He had only semi-quantitative success because of limitations in hardware, but, as a result of his fortuitous use of an unsteady state formulation he discovered that the process was oscillatory even before it was revealed to be so by his experimental measurements.

Using conventional algorithms, David Hellums in 1960 succeeded in obtaining convergent solutions that matched Martini's quasi-one-dimensional but oscillatory experimental results.

Jim Wilkes in 1963 extended the scope of the computations to predict the transient, truly two-dimensional motion in a rectangular channel by introducing the stream function and the vorticity as variables.

Mike Samuels in 1967 extended these computations to predict the criteria for instablity by introducing a "false-transient" term for the stream-function equation, thereby rendering it parabolic.

In the same year Khalid Aziz and David Hellums at Rice University carried out the first three-dimensional calculations for confined natural convection by utilizing the representation of George Hirasaki in terms of "vector potential".

Humbert H.-S. Chu in 1976 calculated the effects of heater-size and location on the energy requirement and temperature distribution in living or working space with losses to the surroundings, and obtained surprising results of great practical interest. At first, he appeared to have paid a penalty for the unorthodox use of a non-conservative finite-difference formulation, namely a different heat flux entering and leaving the space. However, he then made an important discovery, namely that the difference between the computed values of the entering and exiting fluxes was a far better measure of the rate of convergence than the conventional ones.

In the same year, Hiroyuki Ozoe and his co-workers at Okayama University introduced the isometric representation of particle paths to describe the fluid motion in three-dimensional natural convection as an analog of the stream function in two dimensions and discovered that all paths are concentric double helices, or degeneracies thereof. They confirmed this behavior predicted experimentally.

In 1982, Paul P.-K. Chao developed a program for displaying particle paths in real time long before standard algorithms for this purpose became available.

In 1995, Vicki B. Booker and coworkers at the NEC Corporation in Tsukuba, where she was a visitor, found experimentally that Czochralski crystallization, which is widely used to make pure silicon, is oscillatory, thereby rendering her steady state computations, made at the University of Pennsylvania, invalid. She confirmed this observation computationally by formulating an unsteady-state algorithm that predicted transient behavior but whose long-term execution required computational times beyond that available.

In 1998, Hiroyuki Ozoe and coworkers at Kyushu University subsequently confirmed the existence of oscillations in confined natural convection in low-Prandtl-number fluids even for the simpler geometry and boundary condition of parallel horizontal parallel plates heated from below (*the Bénard problem*). The circulation shifts erratically between hexagonal cells and long cylindrical roll cells.

In 1971, Warren Seider, in the process of numerically modeling the developing flow and transport following the merging of a central jet and an outer annular flow, unexpectedly encountered an instability that was subsequently confirmed experimentally. The computational instability was found to be attributable to the finite thickness of the short inner tube, while, on the other hand, the experimental instability was found to be due to any minute radial perturbation in the flow, such as that caused by a bubble of air on the tube wall or someone walking in the laboratory. He ingeniously avoided the numerical instability by carrying out the computations in time, starting from a condition of no flow, thereby confirming experiments in which all physical perturbations were minimized and the flow remained stable as it developed.

The Marker-and-Cell model of Harlow and coworkers is uniquely applicable for solving the partial differential equations of conservation numerically for unbounded flows. Eddy A. Hasbun, in 1973, utilized it successfully to predict the extrusion of Plexiglas^{R} in three dimensions.

The lesson from these examples is that the prediction or simulation of transport often depends on conceptual innovation, either mathematically or physically.

Simulation

Simulation in the current sense allows us to use correlations for transport to predict complex behavior for the purposes of design and analysis. The advancement and current state of simulation are beyond the scope of the current presentation. However, it seems appropriate to note that this process invokes a hidden risk, namely the possible error due to out-of-date and erroneous correlating equations imbedded in computer packages. Simulation has another almost contradictory role that has been implicit in this presentation, namely the prediction of detailed behavior from "first principles" in order to produce "computed values" to supplement experimental data in support of the construction of correlating equations.

The Influence of AIChE Programming

Industrial and academic attendees have a basic conflict over programming **at** AIChE meetings. Academics generally prefer to present work in progress with the hoped that that the presentation will provoke constructive criticism and perhaps establish informal proprietary rights. They are inclined to entrust the presentation to a graduate student because of the invaluable professional experience, the start in networking, and the encouragement to participate in future meetings. Accordingly they resist the burden of preparing and distributing a document that will soon be out of date. Academic attendees who are not making presentations understand and support this posture.

On the other hand most industrial participants attend the technical presentations hoping to learn something that will help with their current project and appreciate a manuscript to take back to their work place as evidence of their attendance and as proof that it was worth the time and cost. They also would favor a presentation by the faculty member who supervised the work rather than by the graduate student who actually did most of the work. .

In 1970, the Program Committee, hoping to increase industrial attendance promulgated the policy "No paper, no podium." A group of "young Turks" led by Andy Acrivos, among others, struck back. Their efforts led to the formation of the Fundamentals Section of the Program Committee, and a policy of 15-minute presentations with 5 minutes for discussion. This policy produced a revolution in fluid mechanics. I believe it is fair to say that as a result of this new policy the primary frontier in that subject shifted from all fields of engineering in all countries to the Annual AIChE meeting. Unfortunately, the change in format was far less successful in heat transfer as I will subsequently explain.

External Influences

Sources of investment in and support for advances in transport The political, social, and business climate in the USA has had a major impact on research and development in transport over the past century. Before the founding of the AIChE, the study of transport was most highly developed in Europe and its academic institutions. The early studies of transport in the USA were carried out in industry and often in the context of the operation of process equipment rather than research. Prior to World War II, most advances were prompted by new chemical processes and were discovered by the analysis of full-scale or semi-scale operations as well as by laboratory-scale experiments. During the war the same pattern occurred in petroleum refining. Since that time, both industrial research and academic research, the major contributors to advances in transport, have gone through several ups and downs.

At the end of the war the military services, recognizing the major impact that basic research had had on weaponry and other materiel, and awash in appropriations that had not yet been scaled back to peace-time levels, decided to utilize the excess funds for research rather than give them up. I was a beneficiary of such support for nearly a quarter of a century, and found it more enlightened and free of restrictions than any since. Those golden days came to an end in 1973 by virtue of the "Mansfield Amendment", which, provoked by opposition to the Viet Nam War, forbid *DOD* to support nonmilitary (non-mission-oriented) research.

Industrial research laboratories, such as Shell Development, Esso Research and Engineering, and Bell Laboratories, became pre-eminent in the post World War II years and attracted some of the best doctoral graduates in chemical engineering. They made significant fundamental contributions to transport, of which I will mention one that may not be known to all of you, namely the invention of zone refining in 1950, an essential step in the development of transistors, by Bill Pfann, a chemical engineer, at Bell Labs. Beginning about 1960, basic research fell out of favor with the corporate culture and these industrial research centers gradually lost their prominence and ceased their innovations, at least of a fundamental nature.

In the post World War II years, the chemical industry, and particularly the petroleum industry, was a major source of support for graduate work in chemical engineering in the form of fellowships. This unique source of support, which elevated graduate work in chemical above that in all other branches of engineering, was phased out beginning in the 1960s. The proffered reason for this withdrawal was the fulfillment of that need by *NSF* and *NASA*, but the real reason was their own withdrawal from research. .

The direct support of academic research itself in chemical engineering by industry has had its ups and downs and is currently negligible. The first contract research was apparently carried out at the University of Michigan in the 1930s by Walter L. Badger for the Swenson Evaporator Company. Such industrially contracted research grew for a few years but soon faded because of competition from governmental research and proprietary concerns.

A significant event was the founding of the Heat Transfer Research Institute in 1962. Up until that time, the major chemical and petroleum companies sponsored a significant amount of academic research in heat transfer. That sponsorship was promptly terminated along with their own research in that field. Most of the work undertaken by *HTRI* was quite applied, but because it was proprietary, an academic researcher in heat transfer, could no longer know whether or not the work he was contemplating was already completed or in progress. The impact was soon evident. Up to that point, the participation of chemical engineers in the US National and International Heat Transfer Conferences was roughly equal to that of mechanical engineers. Thereafter, it faded to nearly zero. Research in heat transfer by mechanical engineers survived because their work was supported by industries, such as those involved in aircraft and spacecraft, that did not join *HTRI*. One noteworthy exception was the contract research sponsored by Wolverine Tube, at the University of Michigan, which extended about 35 years under the consecutive direction of Alan Foust, Don Katz, and Ed Young, but that project too is now history.

The National Science Foundation came into being in 1950 but was not a major factor in academic research in chemical engineering until 1957, when *Sputnik* prompted panic over the state of research in the USA. *NSF* gradually became the major and essentially only source of finding for research in transport. Then in 1980 that source of support suffered a devastating blow, namely, the directive that all research sponsored by NSF should have a mission aligned with those of the "government" such as that of the *Strategic Defense Initiative*. If enforced rigorously, that directive would have had the effect of replacing the judgment of individual researchers as to what has promise with that of administrators. Even if the administrators of contracts and grants were "the brightest and best", I would still have more faith in the productivity of exploratory research by individual researchers on the topics that they consider to be most promising. The impact of this policy has been aggravated by the lumping of funds into a few large grants on favored topics. The academic community is necessarily adept at coping with such policies, but innovation suffers. Do you believe that it is currently possible to obtain support for research on some aspect of transport without hiding it in work on energy conversion, the reduction of pollutants, nanotechnology, or biotechnology? Most of the cited advances in transport over the last 20 years have been from unsupported or "bootlegged" research.

Summary

I have presented a few illustrations of our progress over the past century in predicting transport. They are just that – illustrations; I have been highly arbitrary in my choices. Some of that progress consists of the abandonment of familiar concepts – a painful process and one that risks the appearance of criticism of idols of mine as well as yours. I trust that if they were here they would approve. In that regard, I recall W.K. Lewis, in an anecdotal lecture at an AIChE meeting, mentioning that the *Lewis number* commemorated his worst conceptual error.

Most of the advances that I have described today originated in academic research and most of the remaining ones in industrial research. However, I was able to identify them only because they have achieved sufficient recognition, acclaim, and certification to appear in textbooks and handbooks. The ultimate certification is their adoption for design, operation, and analysis, but that is difficult to quantify, except perhaps their "appearance" in computational packages. A few advances were mentioned that have a more limited but important role, namely improvement in understanding, and a few that had an even more limited objective, namely resolving some technical or scientific uncertainty.

Conclusions

Over the First Century of the *AIChE*, the most obvious improvement in the prediction of transport is a consequence of the development of powerful computer hardware and user-friendly software. However, that advance in itself would have had little impact without three others

1) the formulation of theoretically based models for transport, primarily by non-chemical engineers
2) the formulation of special-purpose algorithms, primarily by chemical engineers
3) the interpretation of the computed values in theoretical and algebraic terms, almost wholly by chemical engineers

Two implicit elements should also not be forgotten

1) The accuracy of numerical solutions depends critically upon the validity of the model and on the convergence and stability of the methodology used for its solution.

2) Trusting models and/or their solutions whose limits of accuracy and validity have not been tested with experimental data is equivalent to believing in the Easter bunny.

The only reliable expressions for the prediction of transport are those that have a theoretical structure **and** have been confirmed by both experimental data and numerical simulations. Some of you who work in process design or operation may dismiss what I have said as mathematically oriented and once removed from practicality. That is a dangerous conclusion. In the early days of the AIChE correlating equations were devised by drawing a straight line through a log-log plot of experimental data. Over the century we have come to realize that expressions so-derived are almost certainly in error functionally, and thereby, outside a narrow range, may be in serious error numerically as well. If you are clinging to any such expressions involving products of arbitrary power-functions, for example the Colburn analogy, you are dangerously out of date.

The principal improvement needed with respect to the prediction of transport in conventional systems is the development of a methodology for the accurate prediction of developing turbulent flow and of convection in developing turbulent flow. The κ-ε model purports to fill this need but it is highly unreliable near the wall, which is the most critical region.

The current lack of interest in and support for research in transport is clearly a consequence in of the expansion of biological processing, a related expansion of interest in product as opposed to process design, and an extension of the scale of processing from the macro to the nano.

Bulk processing of chemical materials will undoubtedly remain an important part of our economy and energy conversion may spur a renewed interest in transport on the macroscale, but I suggest that we take advantage of the opportunity to expand the scope of transport to encompass biotechnology and nanotechnology.

We can take justifiable pride in the advances in transport during the *First Century of the AIChE*. I anticipate that comparable advances that we cannot yet imagine will occur in the *Second Century*.

308b A Brief History of the Navier-Stokes Equations: Room for Improvement

Howard Brenner, MIT, Dept. Chem. Eng., MIT, Cambridge, MA 02139-4307

Text Not Available.

The Role of Computation in Continuum Transport Phenomena

Bruce A. Finlayson
Professor Emeritus
Department of Chemical Engineering
University of Washington

Session 308, Perspectives on Transport Phenomena (TH014)
AIChE meeting, paper 308c, Nov. 18, 2008

Status in 1967 when I started my career

- Perry's Chemical Engineers' Handbook, 4th Ed. (1963) (nothing in earlier editions):
  - To solve ODEs: Euler, Adams, simple Runge-Kutta methods
  - To solve PDEs: diffusion/conduction steady problems in 2D (finite difference) or unsteady problems in 1D
  - None of this was reflected in Sections on Fluid Flow or Heat Transmission
- Luther, Carnahan and Wilkes, Applied Numerical Methods (1969)
  - Detailed treatment of numerical analysis, but only explicit techniques with specified time steps

Changes in Perry's Handbook

- 5th edition, 1973
  - For PDEs added alternating direction method and Thomas algorithm for solving tri-diagonal matrics (essential for finite difference methods)
- 6th edition, 1984
  - 2/3 page on finite element method, plus fast Fourier transform, splines, least squares, nonlinear regression, multiple regression
  - In fluid flow section, gave contraction losses, laminar entry flow, vortex shedding
  - In heat transmission and mass transfer, still graphical and algebraic

- 7th edition, 1997
  - Better methods for ODEs, errors, implicit
  - Added boundary value problems (BVP), finite difference, finite element, orthogonal collocation, shooting methods
  - In fluid flow section, more recognition of numerical results: laminar entry flow, sudden contraction, vortex shedding, k-epsilon turbulent models, LES, DNS
  - In heat and mass transfer, nothing
- 8th edition, 2008
  - Stiffness for ODEs
  - Molecular dynamics
  - BVP using spreadsheets and the finite difference method
  - Finite volume methods for PDEs
  - In fluid flow section, mention of numerical results for power law fluids (1978 papers) and viscoelastic fluids (1987 papers)
  - In heat and mass transfer, some linear algebra in radiation section

Numerical Analysis is now used to solve problems ranging from the orientation of nanoparticles to predicting global climate change.

It wasn't always that way.

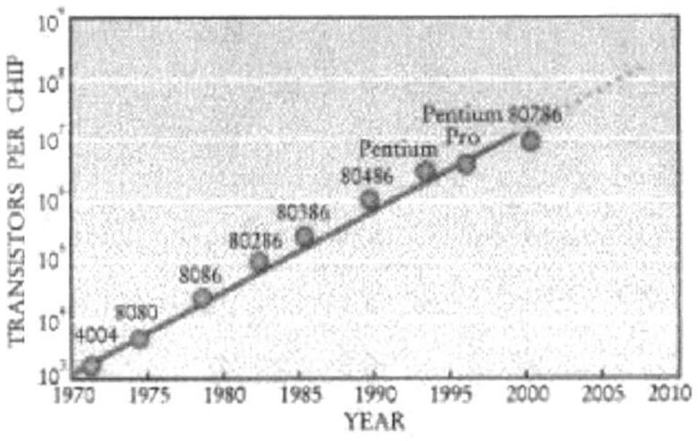

Physics Today, Jan. 2000, p. 40

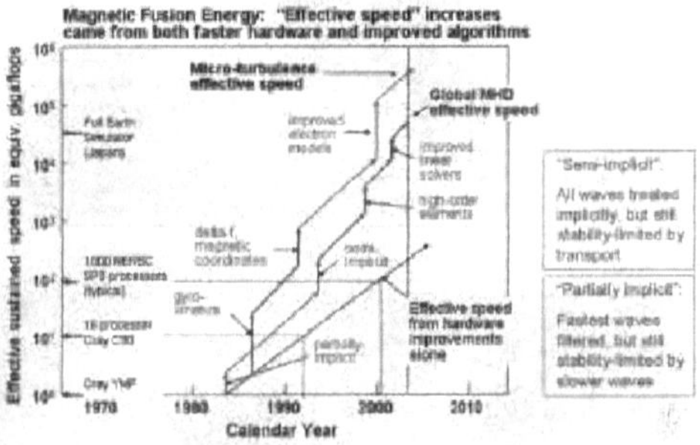

Algorithms help, too! MHD Simulations, faster hardware and improved algorithms, SIAM Newsletter

Numerical Methods for Stiff ODEs

- Runge-Kutta methods existed with error control and automatic step-size adjustment.
- Most engineers used Crank-Nicolson methods, but had to guess a stable step size.

- Gear, 1971; Hindmarsh, 1975, GEARB, later LSODE
  - –When different time constants are important - you want to resolve something occurring on a fast time scale but need to do so over a long time - explicit (RK) methods take a long time.
  - –Implicit methods can be 1000 times faster.
  - –Gear's method allowed for automatic step size adjusment, automatic change of order if that was useful, and basically automatic solution of ordinary differential equations (IVP)

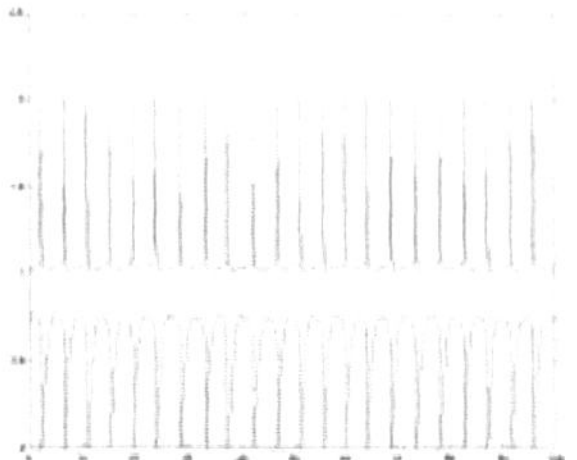

Stirred tank reactor example with a limit cycle

But, the methods are useful for partial differential equations, too!

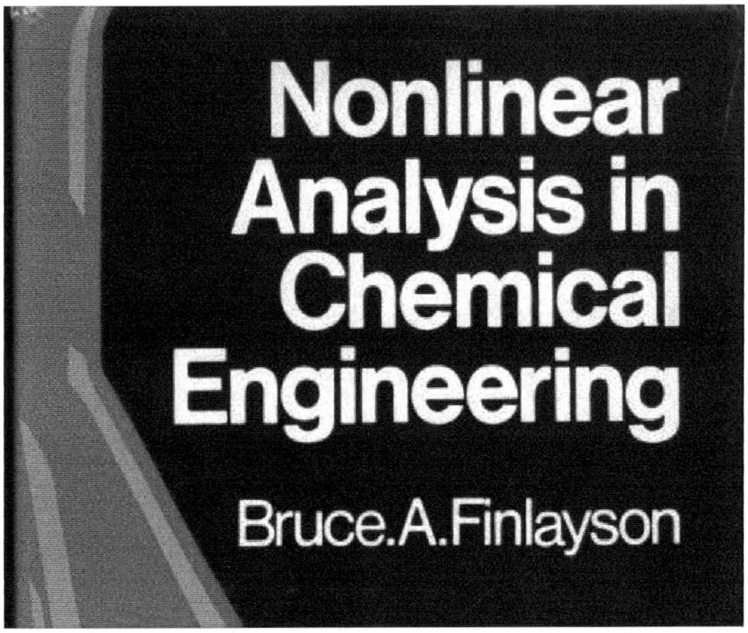

1980 - orthogonal collocation, finite difference, finite element, with programs (still available at www.ravennapark.com)

Orthogonal Collocation - a good idea

Lanczo, 1938 - collocation method with orthogonal polynomials

Villadsen and Stewart, 1967 - solved in terms of value at collocation nodes rather than coefficients - the programming is much simpler

$$c(y,t) = \sum_{i=1}^{N+2} a_i(t) P_{i-1}(y)$$

$$\frac{\partial c}{\partial t} = D \frac{\partial^2 c}{\partial y^2} - \phi^2 R(c) \quad \text{becomes} \quad \frac{dc_j}{dt} = \frac{D}{h^2} \sum_{i=1}^{N+2} B_{ji} c_i - \phi^2 R(c_j)$$

Stiff methods essential for partial differential equations

Depends upon the eigenvalues of the matrix of the Jacobian.

$$\frac{dc_j}{dt} = \frac{D}{h^2}\sum_{i=1}^{N+2} B_{ji}c_i - \phi^2 R(c_j), \quad \left| \frac{D}{h^2} B_{ji} - \phi^2 \frac{\partial R(c_j)}{\partial c_i} - \lambda \delta_{ji} \right| = 0$$

For a diffusion problem, one eigenvalue is due to the problem (is small) and the other is due to the method (and is big).

$As\ N \to \infty\ or\ h \to 0$, the largest $|\lambda|$ gets bigger.

The more accurate your model, the stiffer the problem.

Application to catalytic converter

Involves unsteady heat and mass transport with a complicated rate expression, perhaps eased by occurring in a thin layer of catalyst. The problem may be only one-dimensional, but it must be solved thousands of times in a simulation, even if in steady state. The solid heat capacity makes the time scales very different. Orthogonal collocation models were "4 to 40 times faster (Chem. Eng. J. **1**, 327 (1970).

$$\varepsilon \frac{\partial c}{\partial t} = \frac{D_e}{r^2}\frac{\partial}{\partial r}(r^2 \frac{\partial c}{\partial r}) - kR(c,T)$$

$$\left(\varepsilon \rho C_{pf} + (1-\varepsilon)\rho_s C_{ps}\right)\frac{\partial T}{\partial t} = \frac{k_e}{r^2}\frac{\partial}{\partial r}(r^2 \frac{\partial T}{\partial r}) + (-\Delta H_{rxn})kR(c,T)$$

Catalytic Converter

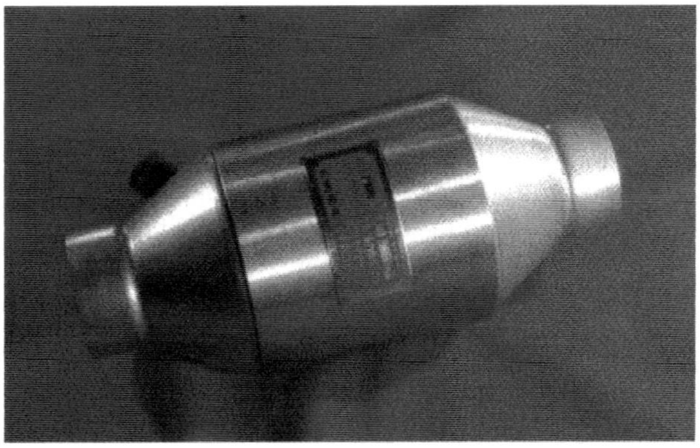

Phenomena included:

Chemical reaction

Flow

Axial conduction of heat

Diffusion

Geometry

What is the importance of the shape of channel?

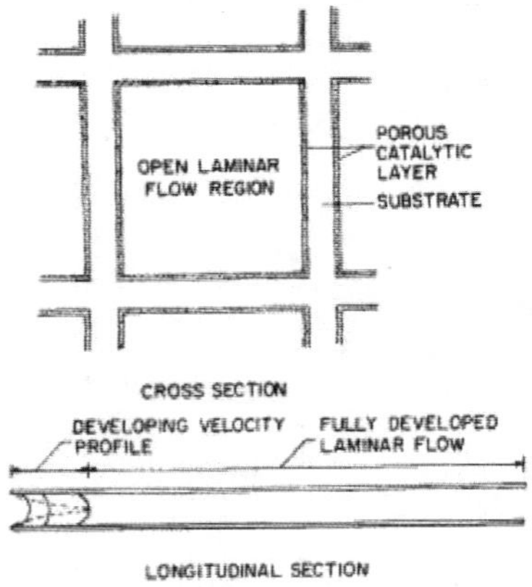

Fig. 1. Cross section and longitudinal section of the square cell.

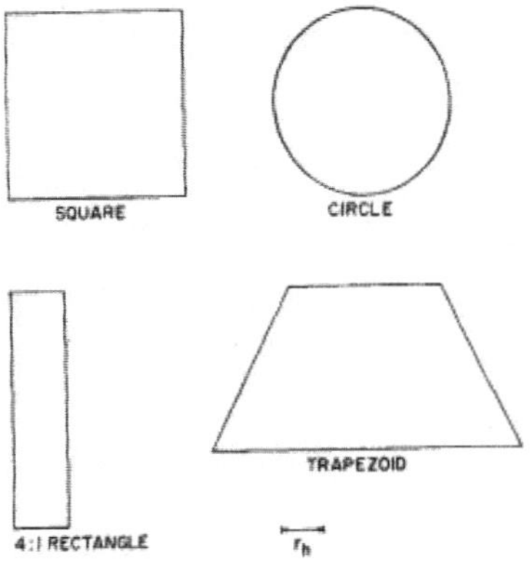

Fig. 2. Cross sections used in this study. Compared at constant r_h.

Page 344 March, 1976 AIChE Journal (Vol. 22, No. 2)

Model I-A is lumped
Model II-A is distributed, using orthogonal collocation on finite elements

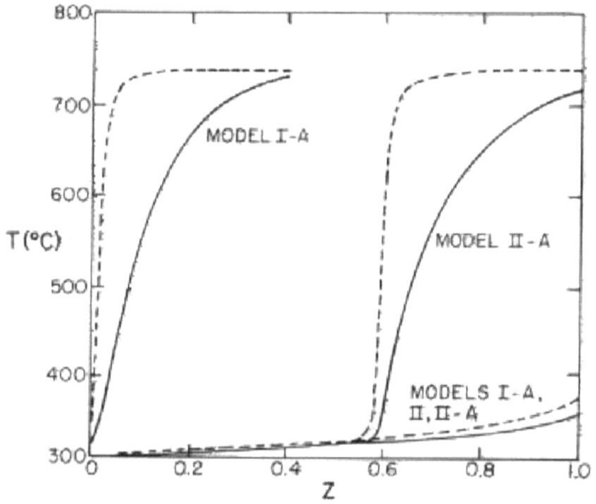

Fig. 8. Steady state model predictions for square geometry. Inlet T = 315°C, y_{CO} = 4%, $\hat{G}$ = 0.0151 m^3/s, D^f_{CO}/D^s_{CO} = 70, ζ = 0.0254 mm. - - - - - Solid temperature, ———— Fluid average temperature. $P2$ kinetics, model I-A: Nu = Sh = 4.0.

Finite Element Method

Began in Civil Engineering for structural problems. The finite elements were beams and rods. It solved the same kind of problems done in Physics 101, except in more complicated structures. Then it was expanded to differential equations.

The dependent variable was expanded in known functions.

$$c(x) = \sum_{i=1}^{N+2} a_i P_{i-1}(x)$$

Key ideas in Finite Element Method

- Cover domain with small triangles or rectangles, or their 3D equivalents.
- Approximate the solution on that triangle using low order polynomials.
- Use Galerkin method to find solution at nodal points.
- Can use higher order polynomials.
- Requires lots of memory, fast computers.

The function x^2 exp(y-0.5) looks like this when plotted:

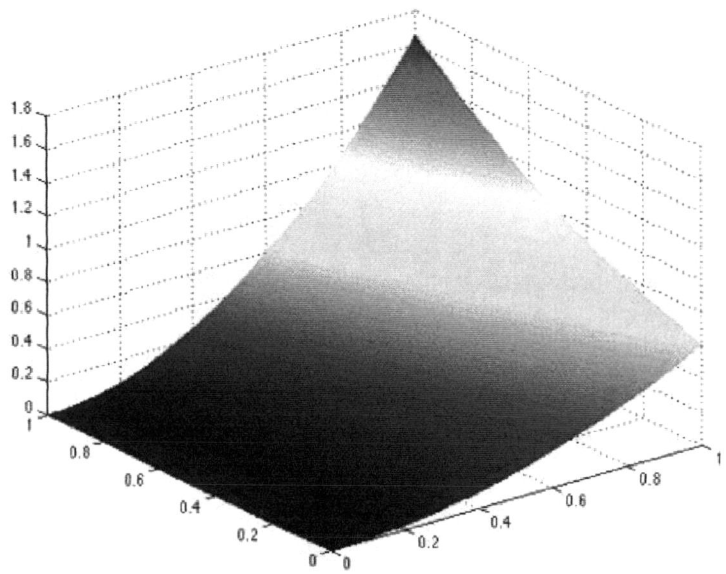

Here is what we expect in a contour plot of the function:

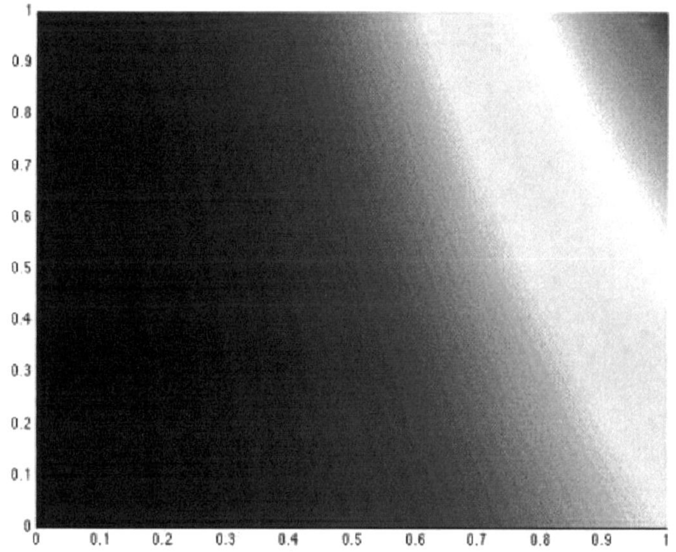

With square elements with one value: N = 4, 8, 16, and 32:

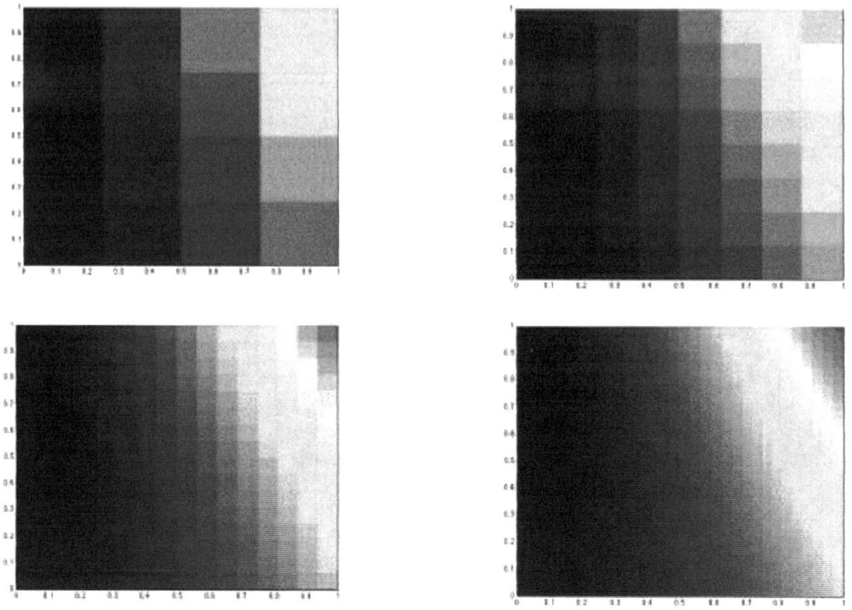

Let functions in the block be bilinear functions of u and v.

- N1 = (1 - u) (1 - v)
- N2 = u (1 - v)
- N3 = u v
- N4 = (1 - u) v
- For example:
- N3(1,1) = 1; N3(0,1) = N3

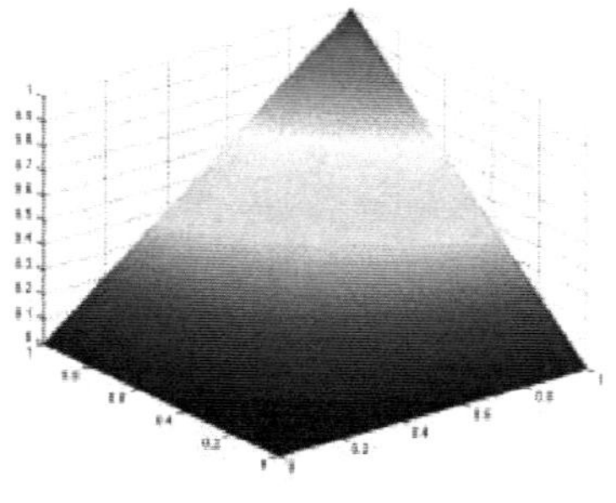

Compare constant interpolation on finite elements with bilinear interpolation on finite elements.

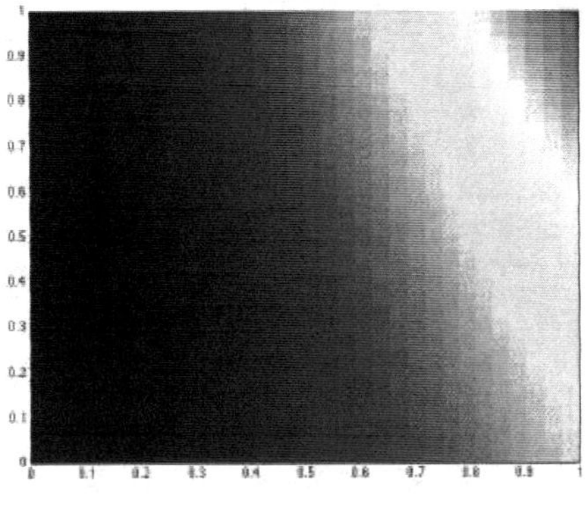

Constant interpolation with 32x32 = 1024 blocks.

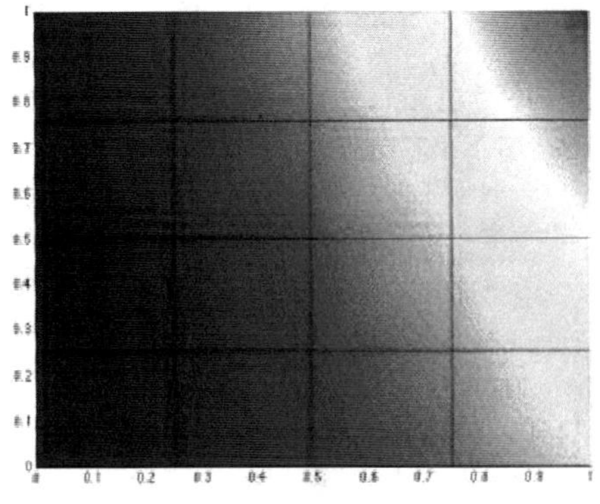

Bilinear interpolation with 4x4 = 16 blocks.

Hole Pressure Problem, Nancy Jackson, 1982

Examined hole pressure for viscoelastic fluids and learned that all four assumptions made by Lodge to relate to the first normal stress difference were wrong, but they all averaged out.

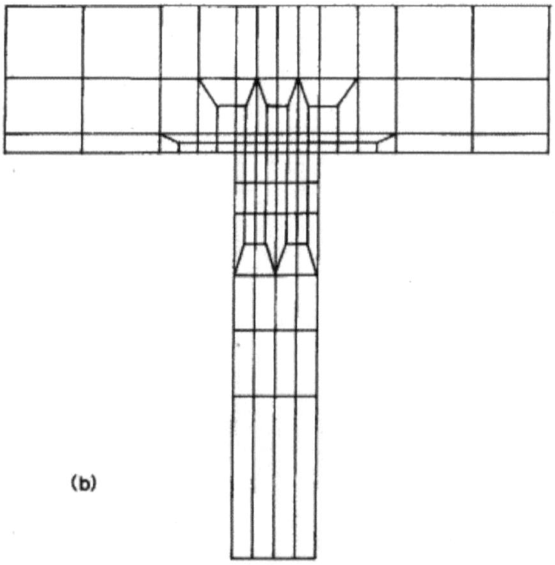

Fig. 3. Finite element mesh: (a) 29 elements; (b) 110 elements.

J. Non-Newt. Fluid Mech. **10** 71 (1982).

Three-dimensional hole pressure
(work done by junior Stephanie Yuen, 2007)

Comparing 2D and 3D calculations. Hole pressure is used in rheology to measure the first normal stress difference.

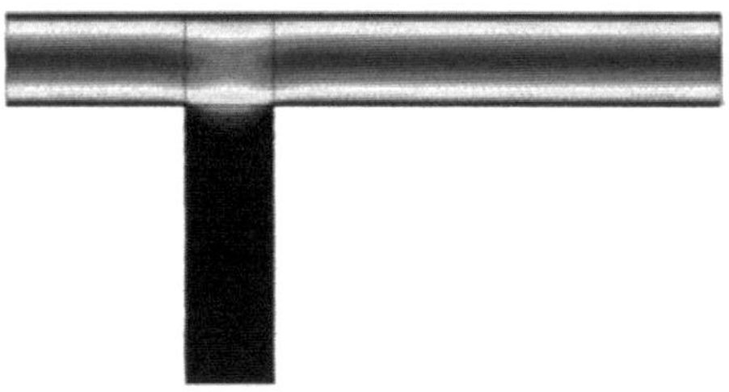

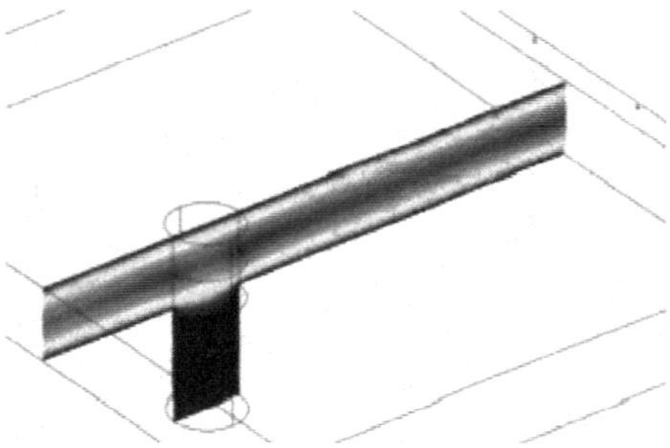

Equations for Viscoelastic Fluid

$$Re\mathbf{v} \bullet \nabla \mathbf{v} = -\nabla p + \nabla \bullet \tau$$

$$\nabla \bullet \mathbf{v} = 0$$

Newtonian Fluid:

$$\boldsymbol{\tau} = \eta\,\mathbf{d}, \;\; \mathbf{d} \equiv \nabla \mathbf{v} + \nabla \mathbf{v}^T$$

Maxwell Model (η, λ constant), White-Metzner Model (η, λ vary with shear rate) :

$$\boldsymbol{\tau} + \lambda\left[\mathbf{v} \bullet \nabla \boldsymbol{\tau} - \nabla \mathbf{v}^T \bullet \boldsymbol{\tau} - \boldsymbol{\tau} \bullet \nabla \mathbf{v}\right] = \eta\,\mathbf{d}$$

Phan-Thien-Tanner Model:

$$\boldsymbol{\tau} + \lambda\left[\mathbf{v} \bullet \nabla \boldsymbol{\tau} - \nabla \mathbf{v}^T \bullet \boldsymbol{\tau} - \boldsymbol{\tau} \bullet \nabla \mathbf{v}\right] + \varepsilon \frac{\lambda}{\eta} tr(\boldsymbol{\tau})\boldsymbol{\tau} = \eta\,\mathbf{d}$$

Stick Slip

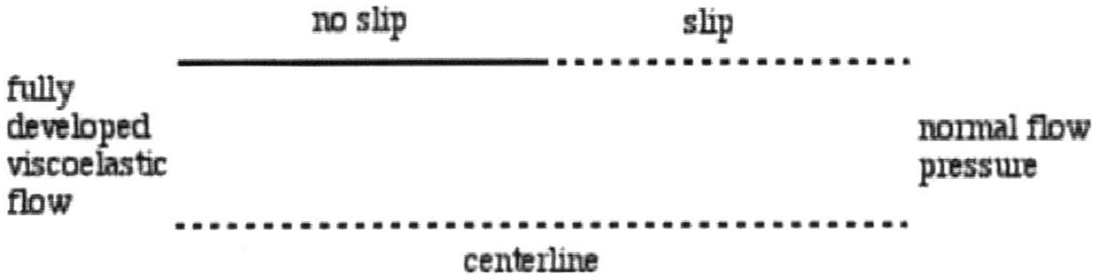

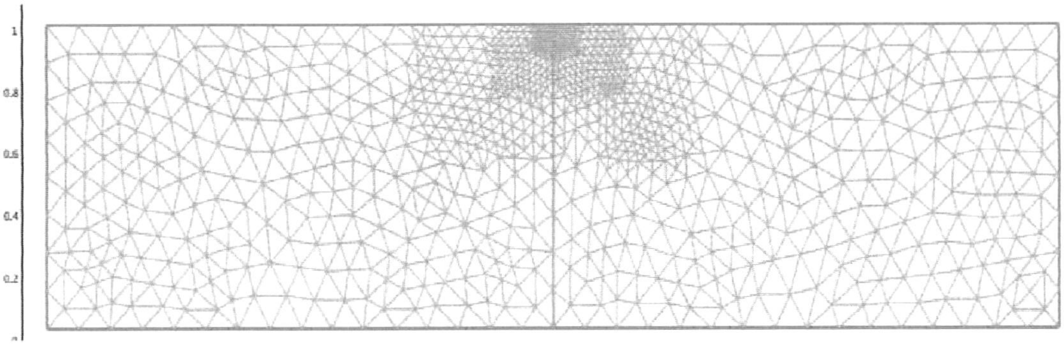

Stick Slip, Standard Method, We = 0:0.01:0.1

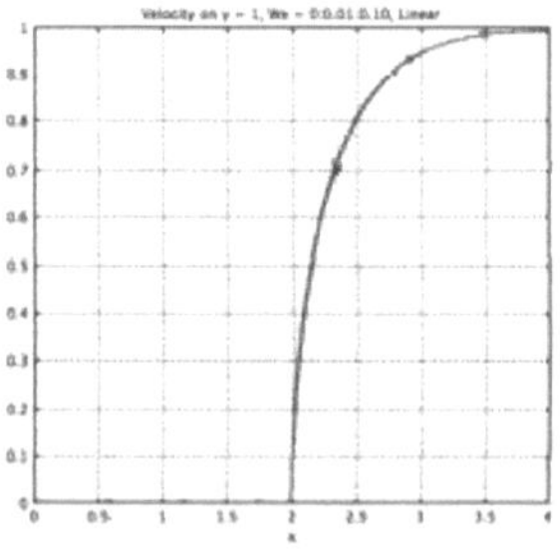

velocity

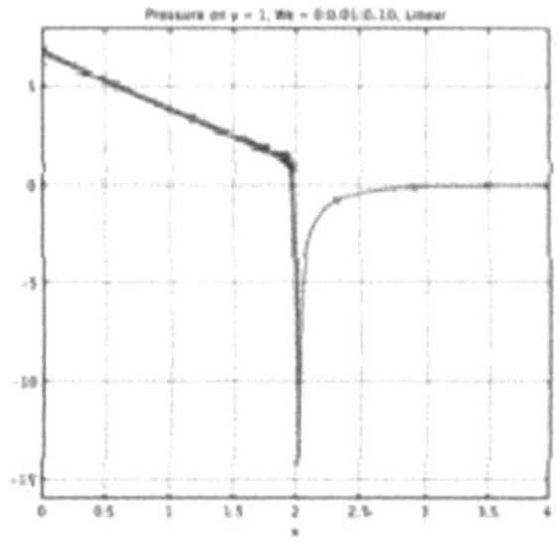

pressure

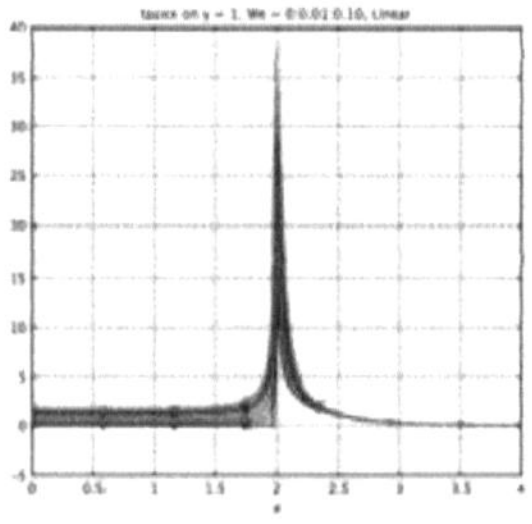

xx stress

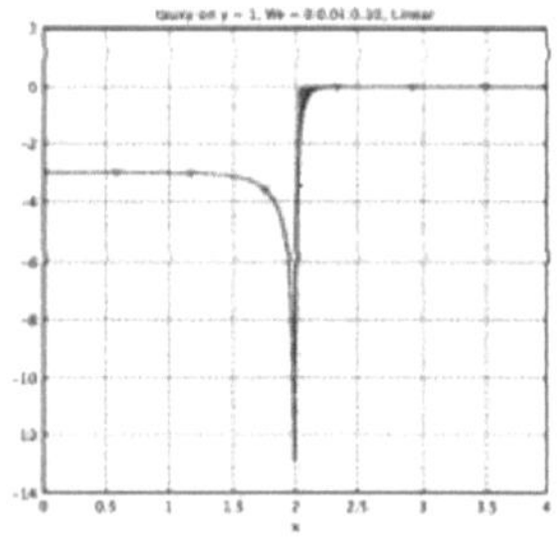

xy shear stress

Stick Slip with DEVSS Method

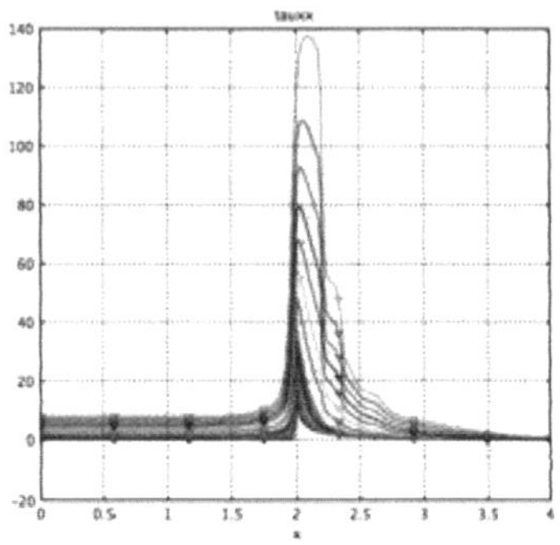

Maxwell fluid,

We = 0:0.05:0.45

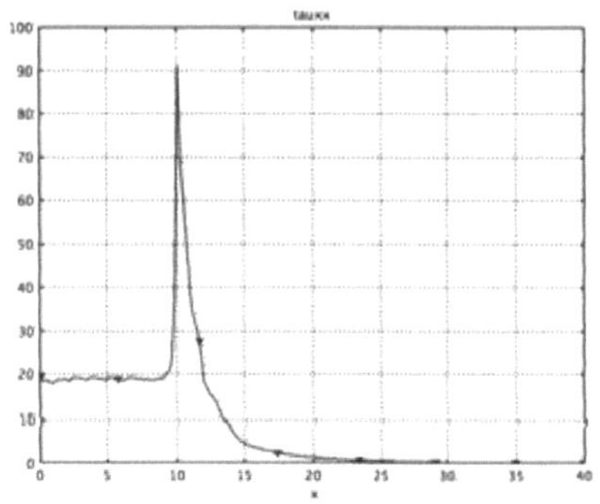

PTT fluid, We = 5

Differential-Elastic-Viscous-Split-Stress (DEVSS)

Guenette, R. and M. Fortin, *J. Non-Newtonian Fluid Mech.* **60** 27 (1995)

R. G. Owens and T. N. Phillips, *Computational Rheology*, Imperial College Press (2002)

Comparison to Experiment

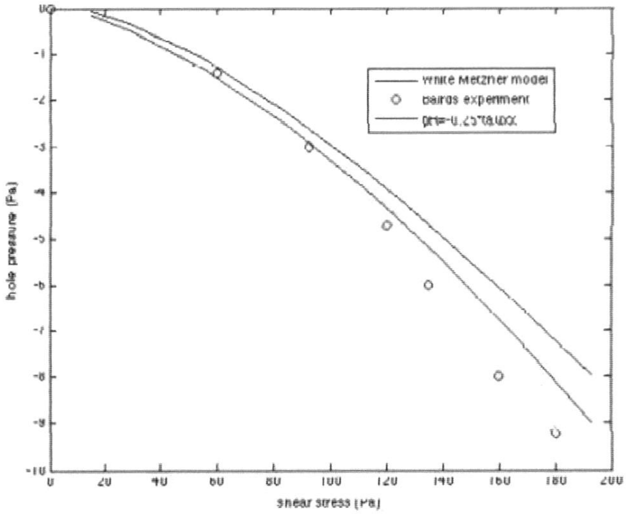

Ref: D. G. Baird, *J. Appl. Poly. Sci.* **20** 3155 (1976)

N. R. Jackson and B. A. Finlayson, *J. Non-Newt. Fluid Mech.* **10** 71 (1982)

Convective Instability, Michael Harrison (2003)
Heat transfer between flat plates, heated from below

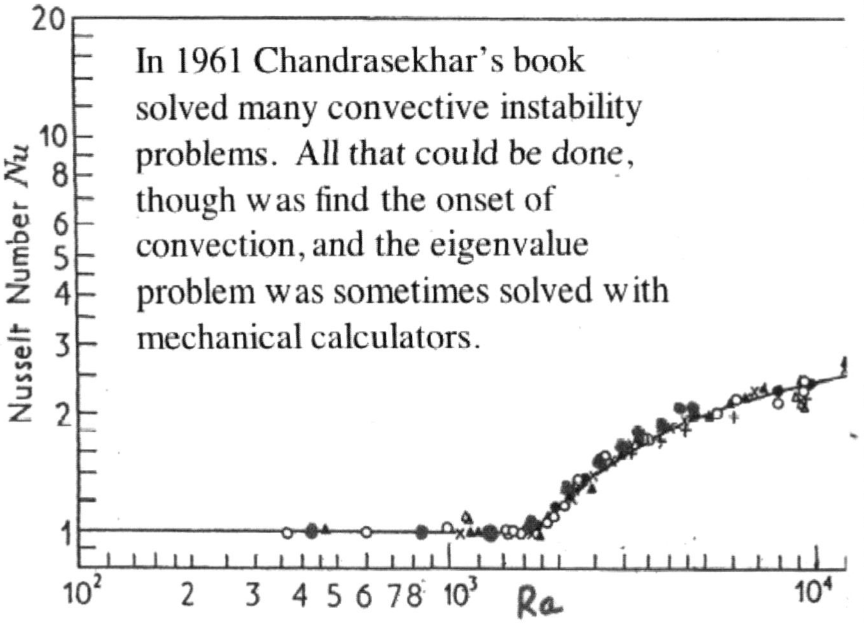

Trapping of DNA using thermal diffusion, Pawel Drapala (2004)

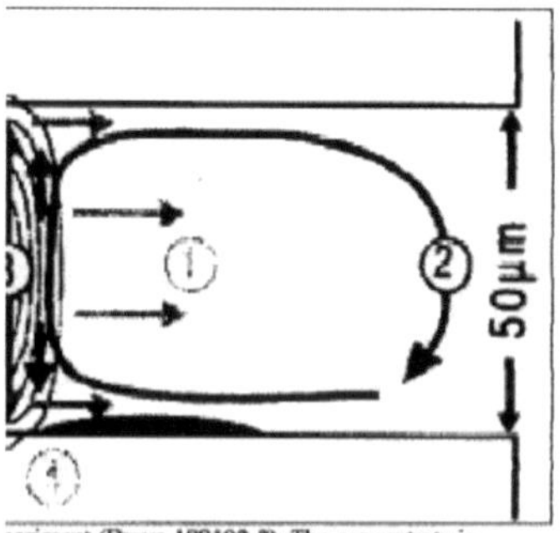

periment (Braun 188103-2). The concentrate is
2), and pilled on the bottom of the container (4).

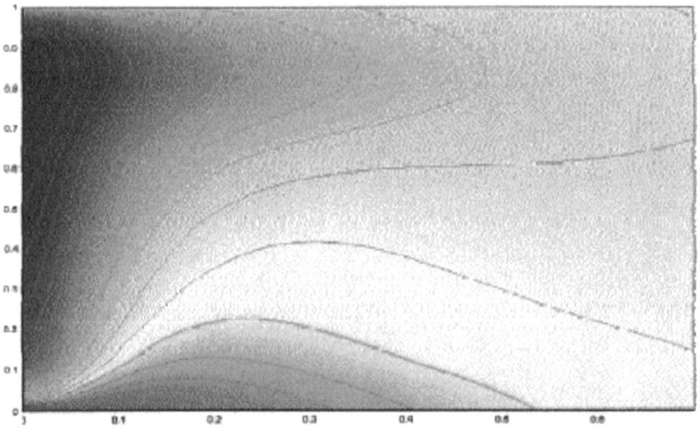

Figure 9. FEMLAB's concentration profile at a final transient value of 5 seconds.

Patterned after experiments by Braun and Libchaber, *Phy. Rev. Letters* **89** 188103 (2002).

Determine Pressure Drop Coefficients for Slow Flow (to mimic those available for turbulent flow)

$$\frac{\Delta p\, x_s}{\eta v_s} = K$$

Table III. Coefficient K_L for contractions and expansions for *Re* negligbily small

	Picture	K_L
2:1 pipe/planar		7.3/3.1
3:1 pipe/planar		8.6/4.1
4:1 pipe/planar		9.0/4.5
45 degrees tapered, planar, 3:1	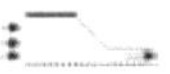	4.9
28.07 degrees tapered, planar, 3:1		10.8
3:1 square (quarter of the geometry)		8.1

v_s = average velocity, x_s = thickness or diameter, both in the small section

Table in Ch. 8, "Micro-component flow characterization," Koch, Vanden Bussche, Chrisman (ed), Wiley (2007). The chapter has 11 authors, 10 UW undergraduates plus Finlayson.

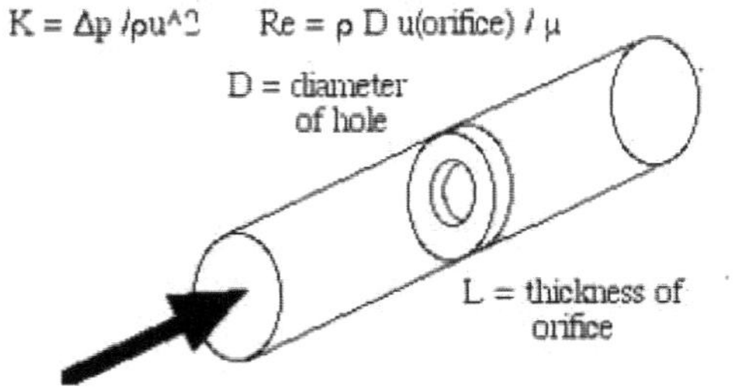

Later, undergraduates could solve harder problems using Comsol Multiphysics (FEMLAB).

Kusmanto, Jacobsen, and Finlayson, Phys. Fluids **16** 4129 (2004)

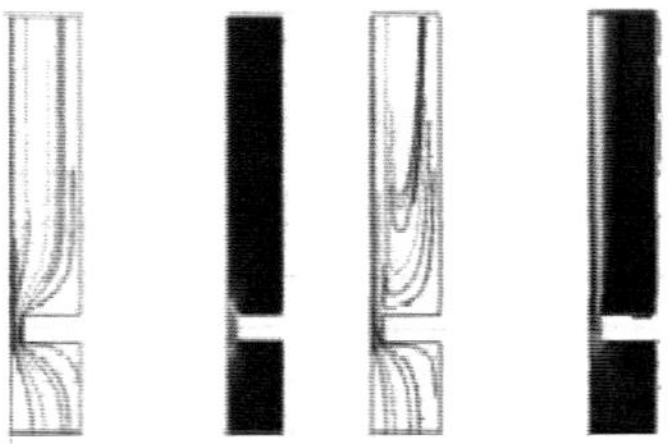

Streamlines and pressure profiles for Re = 0 (left) and 316 (right)

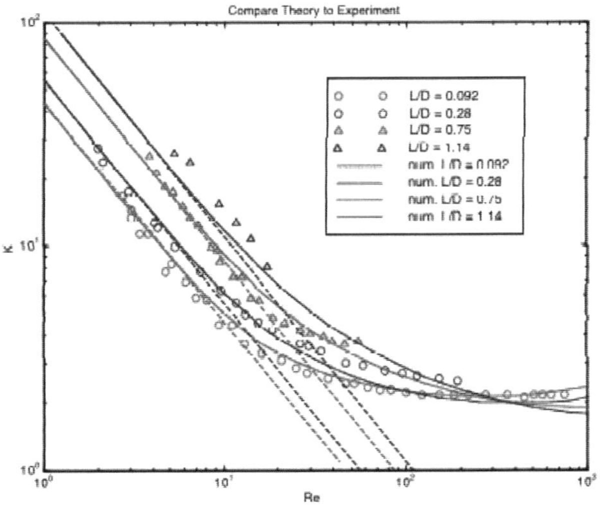

H-sensor - used to separate chemicals by diffusion (solutions by Krassen Ratchev, 2008)

$D = 10^{-9}\ m^2/s$

$D = 10^{-11}\ m^2/s$

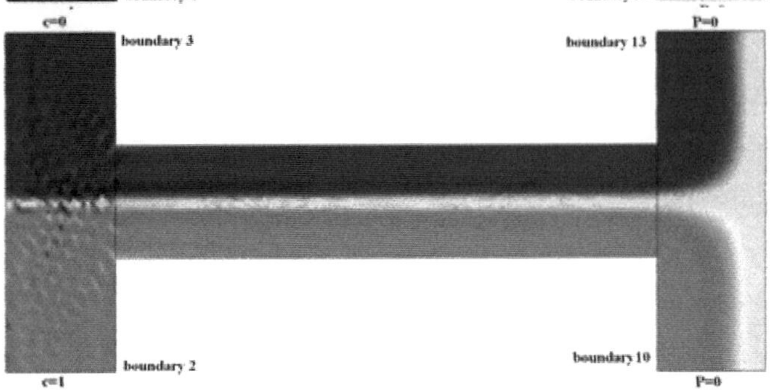

Mixing in a Serpentine Microfluidic Mixer

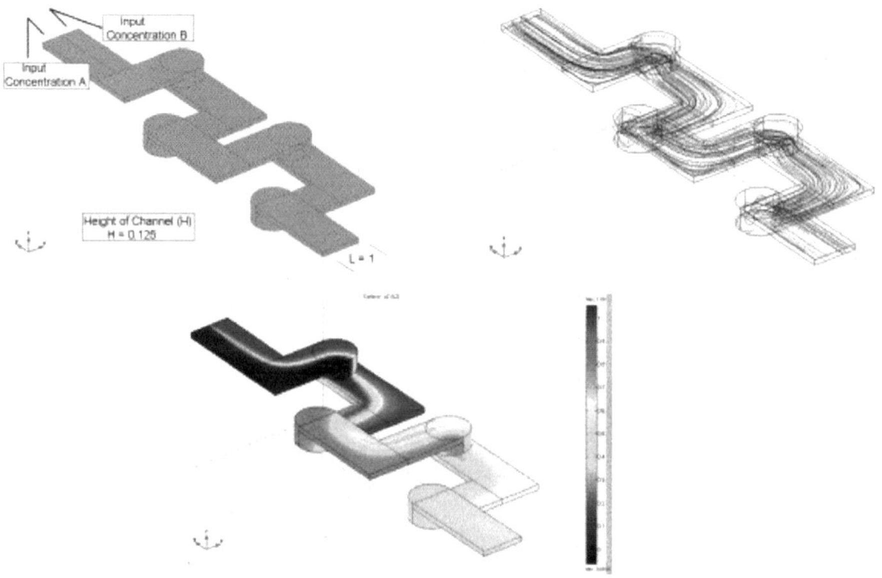

Published in Neils, Tyree, Finlayson, Folch, *Lab-on-a-Chip* **4** 342 (2004)

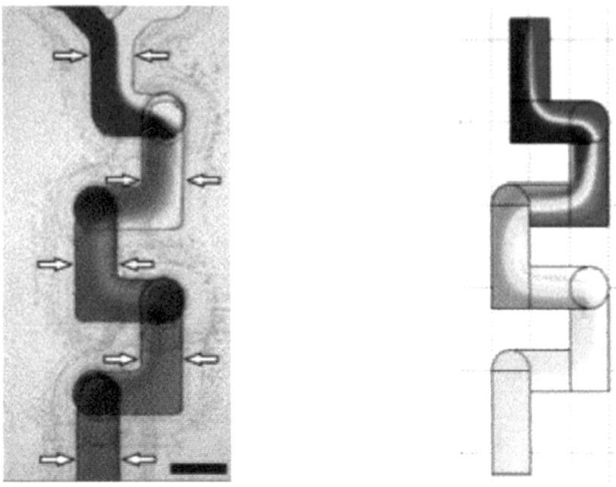

Figure 8. Comparison with experiment (6)

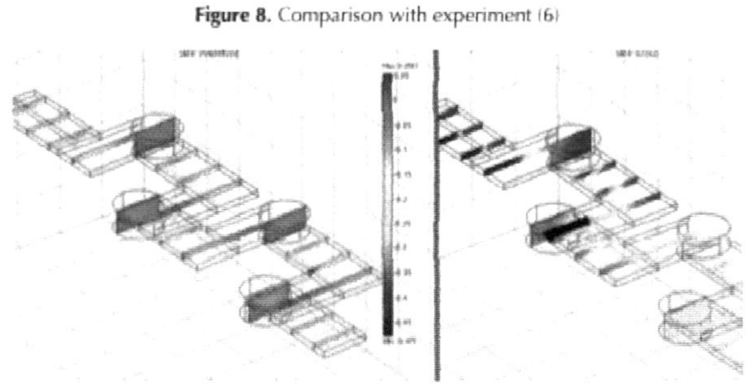

Figure 9. Velocity profiles and concentration profiles inside serpentine mixer

For Re = 1 or so, the flow problem is easy. But, the Peclet number can be large (2000). Then the mesh for the concentration problem has to be refined significantly. Comsol allows solution of the flow problem and the convective diffusion problem on different meshes, thus speeding up the solution time.

Mixing in a Three-dimensional T
(work done by junior Daniel Kress)

$$c_{\text{mixing cup}} = \frac{\int_A c\,\mathbf{u} \bullet d\mathbf{A}}{\int_A \mathbf{u} \bullet d\mathbf{A}}, \quad variance = \frac{\int_A (c - c_{\text{mixing cup}})^2\,\mathbf{u} \bullet d\mathbf{A}}{\int_A \mathbf{u} \bullet d\mathbf{A}}$$

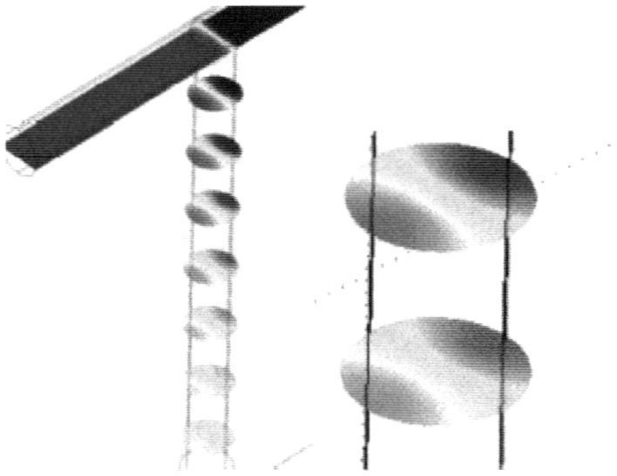

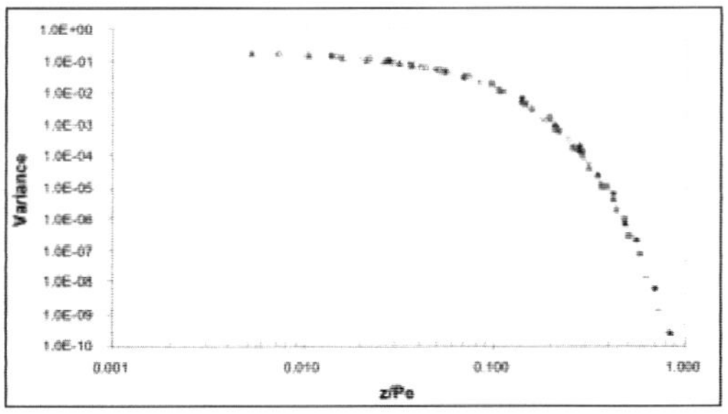

Variance as a function of length in the outlet leg

The work showed that the 3D case followed the same curve as the 2D case (T-sensor).

Mixing in Microfluidic Devices
(11 undergraduate projects)

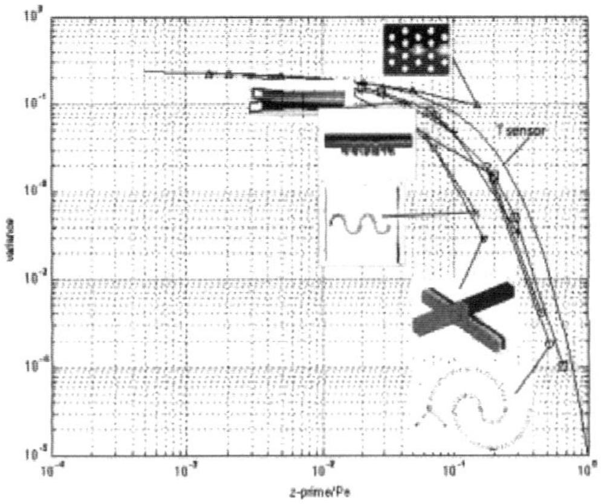

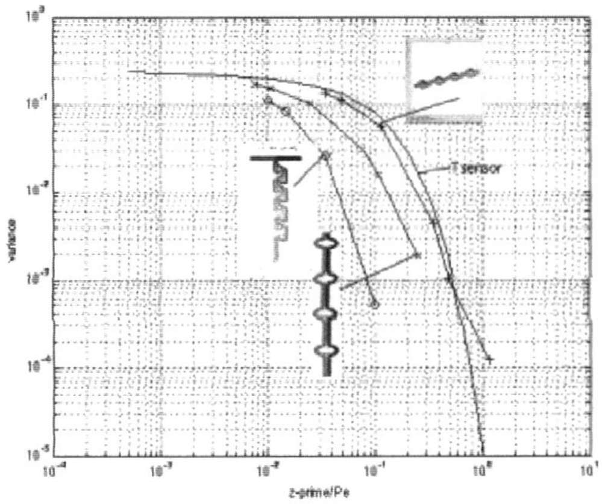

Spin-up of ferrofluid

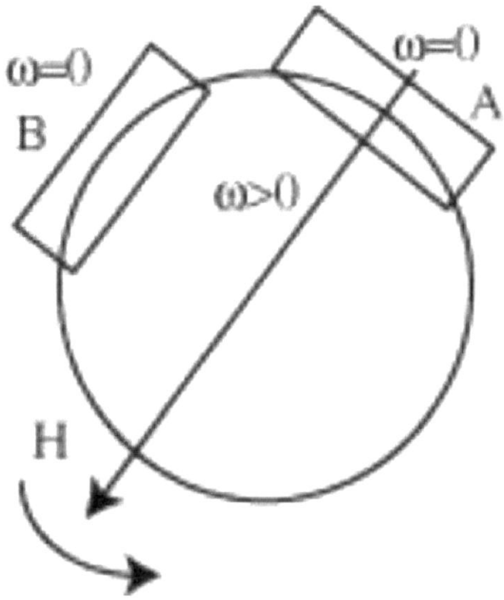

Governing Equations

due to Rosensweig (1985)

Extended Navier-Stokes Equation:

$$\rho \frac{\partial \mathbf{v}}{\partial t} + \rho \mathbf{v} \bullet \nabla \mathbf{v} = -\nabla \mathbf{p} + 2\varsigma \nabla \times \boldsymbol{\omega} + (\eta + \varsigma)\nabla^2 \mathbf{v} + \mu_0 \mathbf{M} \bullet \nabla \mathbf{H}$$

Conservation of internal angular momentum (spin equation):

$$0 = \mu_0 \mathbf{M} \times \mathbf{H} + 2\nabla \times \mathbf{v} - 4\boldsymbol{\omega} + \eta' \nabla^2 \boldsymbol{\omega}$$

Magnetization (Shliomis, 1972):

$$\frac{\partial \mathbf{M}}{\partial t} + \mathbf{v} \bullet \nabla \mathbf{M} = \boldsymbol{\omega} \times \mathbf{M} - \frac{1}{\tau}\left(\mathbf{M} - \mathbf{M_{eq}}\right)$$

Maxwell's Equations for non-conducting fluid:

$$\nabla \bullet \mathbf{B} = \mathbf{0}, \quad \nabla \times \mathbf{H} = 0, \quad \mathbf{B} = \mu_0(\mathbf{H} + \mathbf{M})$$

$$\mathbf{H} = \nabla \phi \qquad \nabla^2 \phi = -\nabla \bullet \mathbf{M}$$

Rotating H and Magnetization

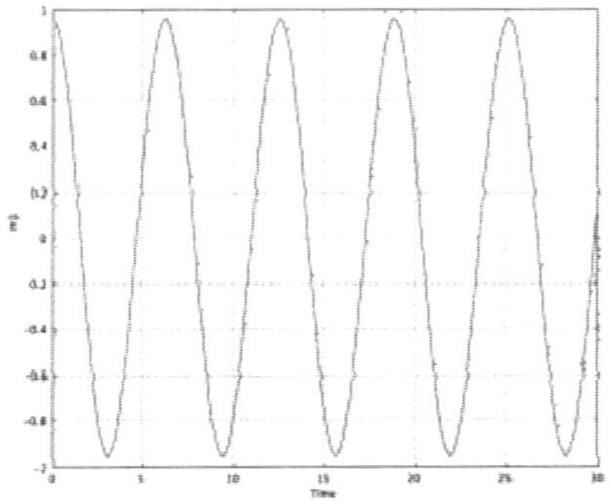

Velocity Field

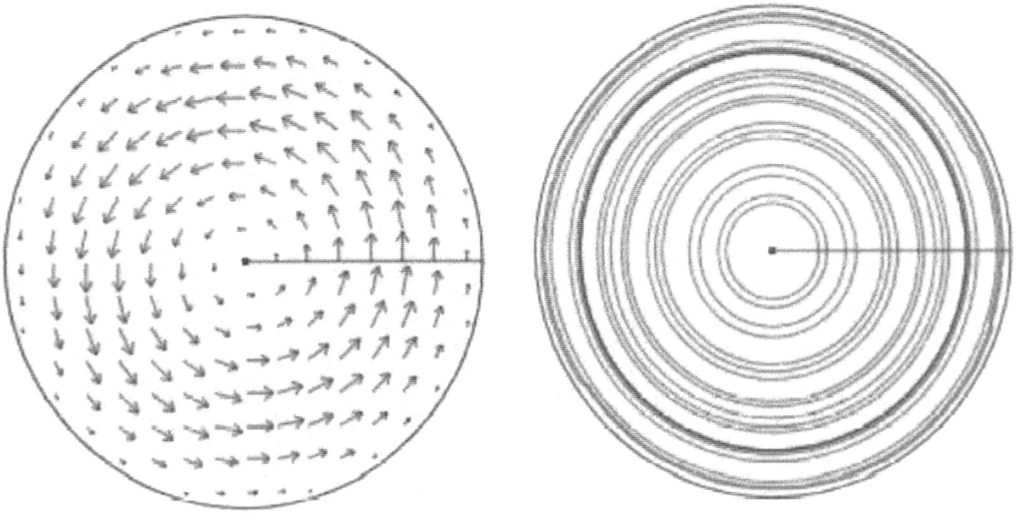

Torque along y = 0

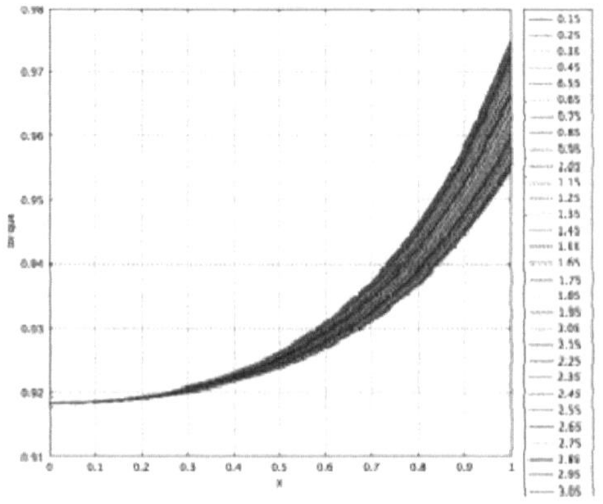

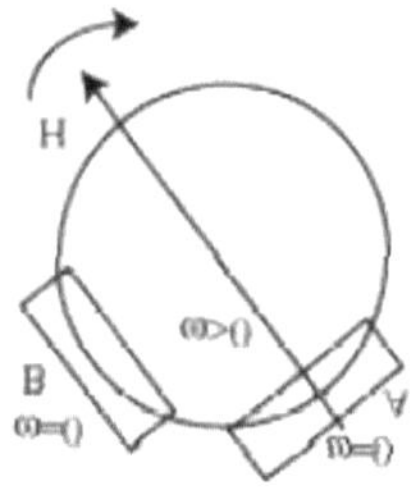

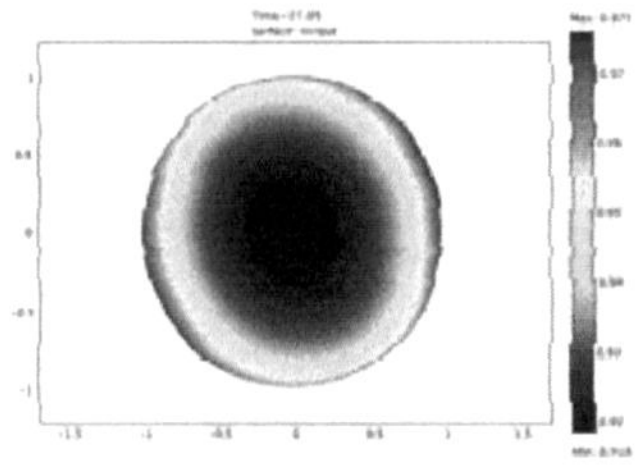

Flow reversal at large H (relative H = 32)

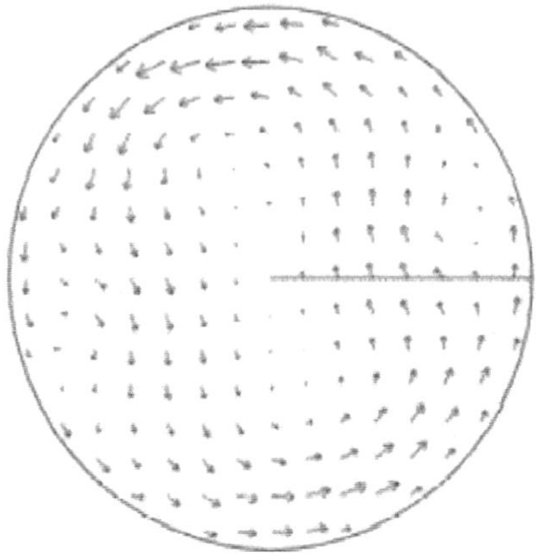

Spin viscosity 10x higher

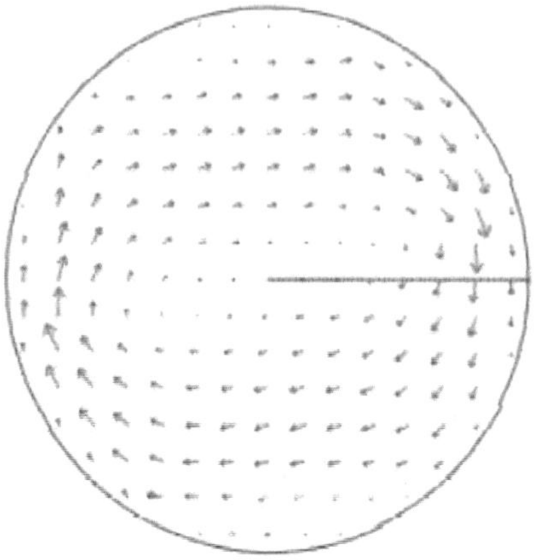

Relative spin viscosity = 1

Heat Transfer to Ferrofluids

Convective instability of ferromagnetic fluids

B. A. Finlayson, *J. Fluid Mech.* **40** 753 (1970)

Using linear stability theory to show when a fluid layer, heated from below, would become unstable.

$$\frac{Ra}{Ra_c}+\frac{N}{N_c}=1,\quad Ra=\frac{\alpha g\beta d^4\rho C}{\upsilon k},\quad N=\frac{\mu_0 K^2\beta^2 d^4\rho C}{\mu k(1+\chi_0)}$$

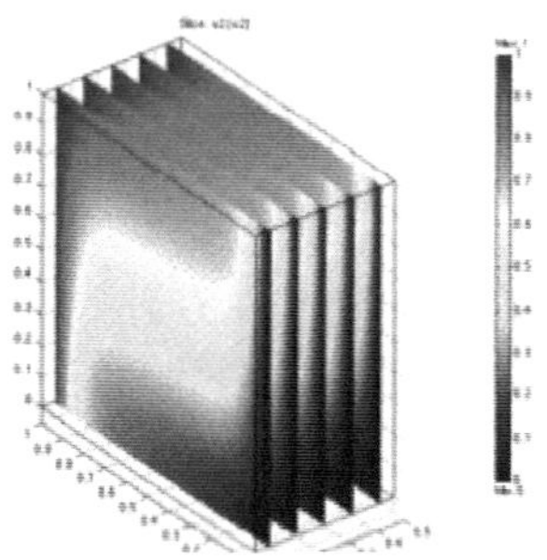

Heat Transfer to Ferrofluid, Suzanne Snyder, *J. Mag. Mag. Mat.* **262** 269 (2003)

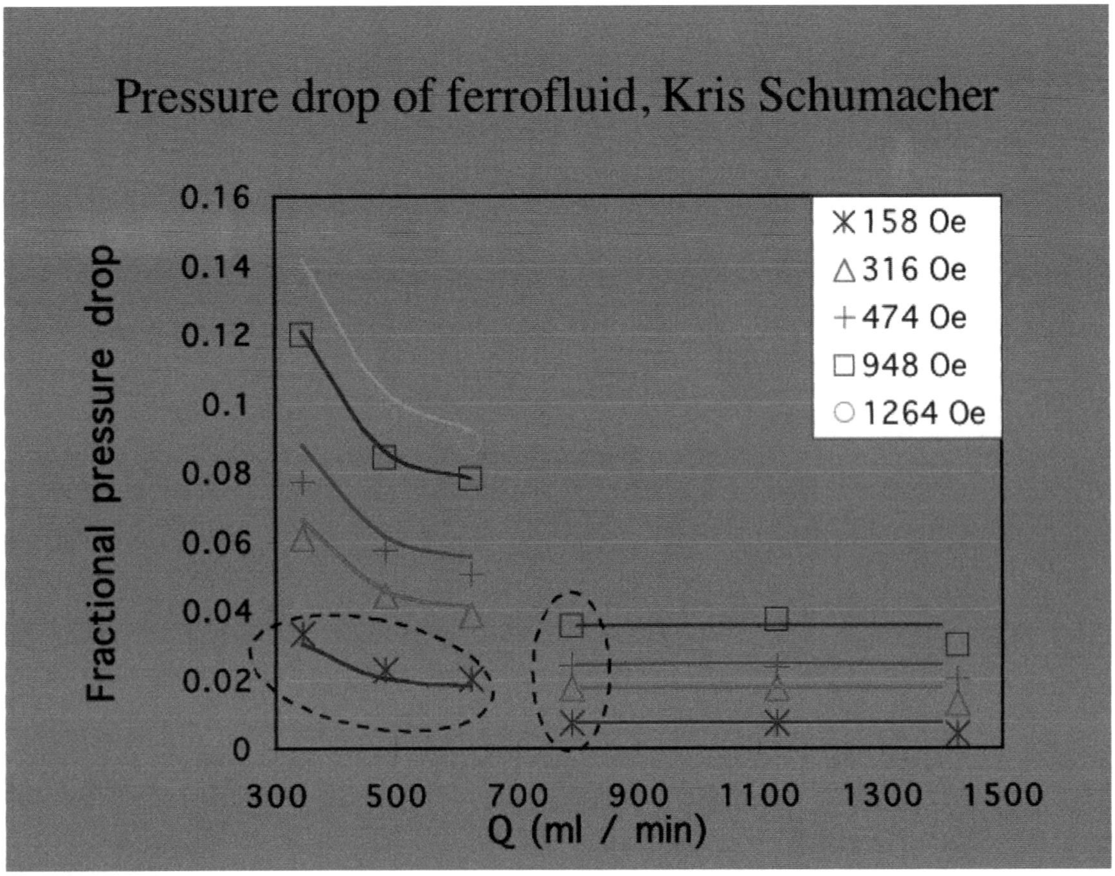

Programs

- Microsoft Excel ®
- MATLAB®
- Aspen Plus ®
- FEMLAB ®

➢Philosophy - students can be good chemical engineers without understanding the details of the numerical analysis.

➢By using modern programs with good GUIs, the most important thing is to check your results.

➢Instead of teaching a small fraction of the class numerical methods, I now teach all the class to use the computer wisely.

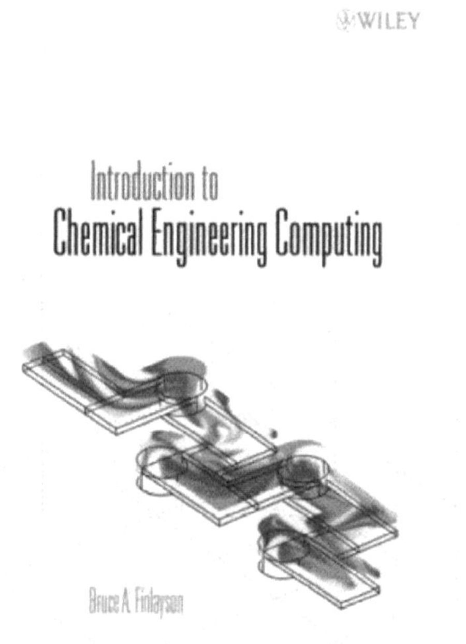

Introduction to Chemical Engineering Computing, transport applications

➢Chemical reactor models with radial dispersion, axial dispersion

➢Catalytic reaction and diffusion

➢One-dimensional transport problems in fluid mechanics, heat and mass transfer

- Newtonian and non-Newtonian
- Pipe flow, steady and start-up
- adsorption

➢Two- and three-dimensional transport problems in fluid mechanics, heat and mass transfer - focused on microfluidics and laminar flow

- Entry flow
- Laminar and turbulent
- Microfludics, high Peclet number
- Temperature effects (viscous dissipation)
- Proper boundary conditions

Steps in Solution

from *Introduction to Chemical Engineering Computing*, Bruce A. Finlayson, Wiley (2006)

- Open Comsol Multiphysics
- Draw domain
- Physics/Subdomain Settings
- Physics/Boundary Settings
- Mesh (Need to solve one problem on at least three meshes, each more refined than the last, to give information about the accuracy.)
- Solve (Can solve multiple equations together or sequentially; can use parametric solver to enhance convergence of difficult non-linear problems.)
- Post-processing (Plot solution, gradients, calculate averages, calculate or plot any expressions you've defined.

Laser Evaporation of a Metal

Westerberg, McClelland, and Finlayson
Int. J. Num. Methods Fluids **26** 637 (1998)

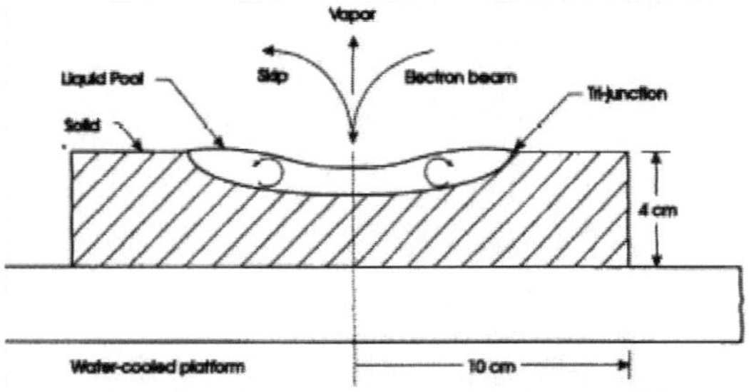

Figure 1. Evaporation of aluminum from a plate on a water-cooled platform.

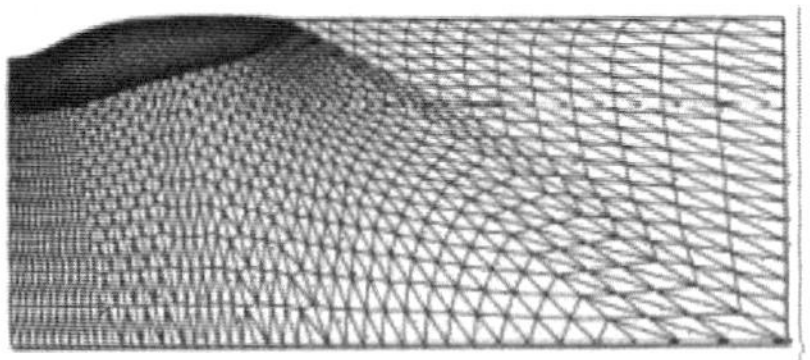

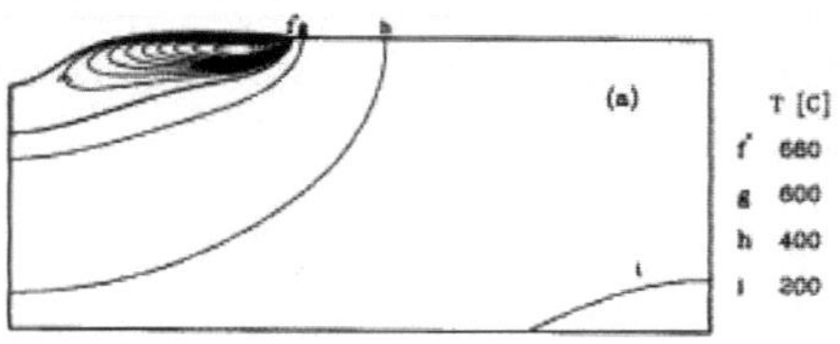

Coating Problems, L. E. Scriven

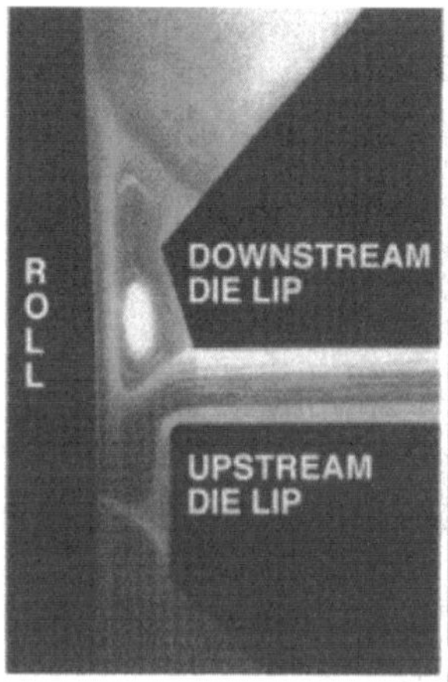

Microvortex made visible with fluorescent dye and hydrogen bubbles in 250-μm-wide gap between coating die lip and substrate moving about 1 m/s. Luigi Sartor and Wieslaw Suszynski.

Fluid-Solid Interaction, Comsol

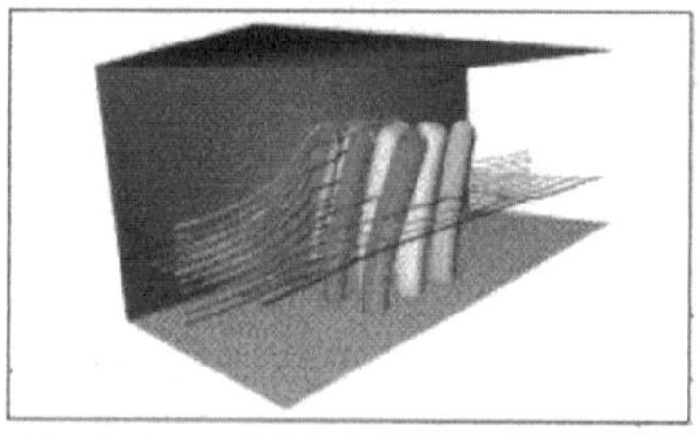

Flow field, pressure, and structural deformation of a set of micro-pillars in a protein sensor calculated using FSI-simulations. The deformation of the pillars depends on the flow field and the amount of adsorbed material. Source: Comsol.

CEP **103** 12 (2007)

Particles in lung

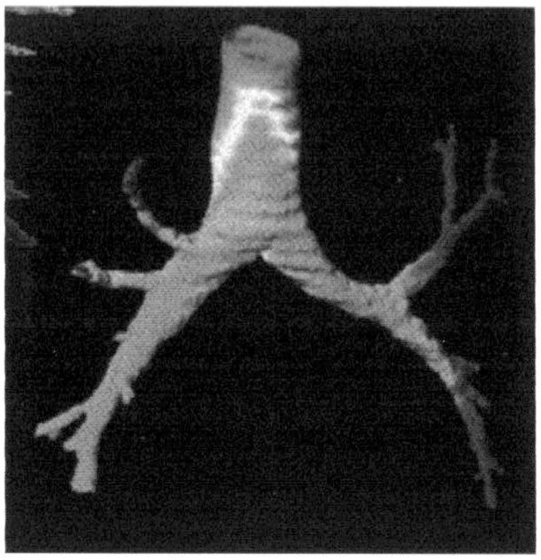

CFX Update, No. 23, p. 26 (2003)

Defects in Materials, Simulations with billions of atoms and fast computers

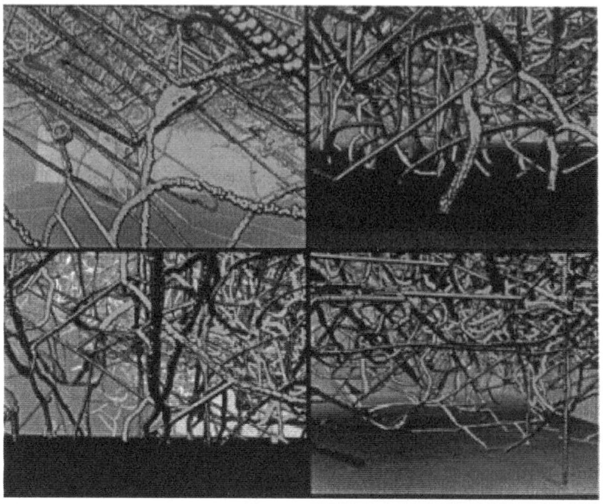

Abraham, et al., *Proc. NAS* **99** 5783 (2002)

Conclusions

- Computer usage in chemical engineering has advanced from non-existent to the solution of very complicated problems.
- Continuum transport problems are being solved routinely using desktop computers, sometimes with commercial software.
- Current tools enable even undergraduates to solve problems in 2D and 3D that were not solvable in 1960.

308d Biological Applications of Transport Phenomena: Converting Mysteries to Puzzles

Edwin N. Lightfoot, Chemical and Biological Engineering, University of Wisconsin--Madison, 1415 Engineering Drive, Madison, WI 53706

All living systems are exceedingly complex transport-reaction networks, and all three of the physical transport processes, along with reaction kinetics, can be involved. The problems we face in dealing with them are normally poorly posed in a mathematical sense, and they are too complex to permit obtaining compete solutions. We shall define such complex problems as mysteries. Description of even relatively simple such systems necessarily requires heuristic approximations, and for those that we encounter here the insoluble mysteries are replaced by approximations taking the form of tractable transport descriptions. We shall refer to these deterministic models as puzzles.

Much of the difficulty in replacing mysteries by puzzles is the almost inconceivably large parameter space in which biological processes take place. However, it is increasingly recognized that the parameter space actually needed to provide useful descriptions can be surprisingly small. This is even true for such complex processes as species evolution and organism development (evo devo).

This short discussion will be limited largely to physiological modeling while at the same time trying to obtain some insight into transport/reaction aspects of evo devo. Fundamental to both efforts is the necessity of reducing the enormous apparent parameter space to the much smaller regions actually needed for the problem of interest. In both we will concentrate on transport related aspects of the overall problem: even tens of thousands of genes are not sufficient by themselves to produce an organism or to facilitate evolution of new species. The importance of non-genetic factors was already recognized by Jacques Monod, but we are just beginning to understand their significance.

Discussion will begin with an overview of universality, pre-disposition, plausibility and self organization, and it will conclude with specific examples. The entire discussion emphasizes the most important characteristic of all evolving systems, whether biological or resulting from human creativity: the phenomenon of emergence. This term is used here to describe the appearance of unexpected results from combining familiar components.

As pointed out by Kirschner and Gerhart (K&G) “understanding of life and life processes is not, however, an immediate goal; it is a long-term evolutionary effort involving many disciplines”. Transport phenomena are among these, and the modeling of physiological transport-reaction systems is itself an evolutionary process.

319a Molecular Simulation as a Foundation for Engineering Models: Impact and Prospects

Keith E. Gubbins, Chemical and Biomolecular Engineering and Center for High Performance Simulation (CHiPS), North Carolina State University, Raleigh, NC 27695-7905

Text Not Available.

319b Progress and Outlook for Monte Carlo Simulation

Doros N. Theodorou, School of Chemical Engineering, National Technical University of Athens, 9 Heroon Polytechniou Street, Zografou Campus, Athens, 157 80, Greece

Since the inception of the Metropolis algorithm in 1953, Monte Carlo methods have been valuable in understanding and predicting the structure and properties of fluids, materials, and biomolecular systems encountered in chemical engineering process and product design. The versatility and wide applicability of these methods is only matched by their mathematical beauty.

Despite the fact that it cannot provide direct dynamical information, Monte Carlo has survived as an indispensable molecular simulation tool. Some reasons for this are the following: (a) It can be adapted readily to sample the probability density of new equilibrium ensembles, judiciously designed to address specific systems and problems. (b) Through creative design of moves that take large strides in configuration space, it can achieve equilibration many orders of magnitude faster than molecular dynamics. (c) It offers the possibility of introducing rigorous bias schemes, or even dispensing with importance sampling altogether, in addressing the thermodynamics of systems with rugged potential energy hypersurfaces. In the last three decades, Chemical Engineers have played a prominent role in inventing and optimizing new Monte Carlo methods. Several of these methods have constituted breakthroughs for the entire molecular simulation community.

Since its early days, Monte Carlo has been indispensable in testing statistical mechanical theories. Today it is viewed as a predictive/design tool in its own right. It is capable of providing quantitative or semiquantitative predictions for the structure, thermodynamic properties, phase diagrams, and free energies of systems of immediate technological interest, based on detailed or coarse-grained molecular models. Implemented in the framework of transition-state theory, it can provide estimates of rate constants for infrequent events. As Kinetic Monte Carlo, it can sample long sequences of such events and thereby track the long-time evolution of diffusion and relaxation processes. It plays a prominent role as part of multiscale modeling/ simulation schemes that bridge the gap between molecular-level interactions and macroscopic properties.

In this presentation we will briefly review some advances in Monte Carlo methods, including Gibbs ensemble simulations, histogram reweighting, parallel tempering, configurational bias and connectivity-altering methods for chain molecules. We will underline the role of Monte Carlo not only in calculating equilibrium ensemble averages, but also in estimating free energy differences and potentials of mean force with respect to specific order parameters, in directly calculating the density of states, and in generating stochastic trajectories for systems whose temporal evolution obeys Poisson statistics. We will discuss insights obtained from Monte Carlo simulations concerning separations in fluid and colloidal systems; the processing properties of molten polymers; the sorption capacity and selectivity of nanoporous materials towards industrial gases; and physicochemical characteristics of pharmaceuticals and biological macromolecules. How Monte Carlo can be combined with other simulation techniques in order to maximize predictive power and minimize computational cost will be discussed.

319c Development of Force Fields: From Empiricism to Quantum Mechanics

J. Ilja Siepmann, Department of Chemistry and Supercomputing Institute, University of Minnesota, 207 Pleasant St. SE, Minneapolis, MN 55455-0431

A review of the evolution of force fields for chemical engineering applications is presented. In particular, it is emphasized how the development of force fields has progressed from parameterizing to properties of non-ideal gases (second virial coefficient and viscosity), to liquid-state properties, to vapor-liquid equilibria, to complete phase diagrams, and, most recently, to data obtained from quantum mechanical calculations. Successes and failures in force field development are discussed.

319d Accuracy of Simulation Compared to Experimental Measurement

Anne M. Chaka, Physics Laboratory, National Institute of Standards and Technology, 100 Bureau Drive Stop 8443, Gaithersburg, MD 20899-8443

Text Not Available.

Building on History: Pathways to Energy Independence and Sustainability through Gasification and Sequestration

Justin "Judd" R. Swift, Office of Fossil Energy, U.S. Department of Energy

1.0 Historical Perspective

For over 150 years, products derived from coal gasification have found wide commercial applications. For example, "town gas" is a product of coal gasification that is composed of 50% hydrogen, with most of the remaining fraction being methane (CH_4) and carbon dioxide (CO_2), and a relatively small percentage of carbon monoxide. In 1816, Baltimore, Maryland, began using "town gas" to provide lighting for residences, streets, and businesses. Baltimore Electric Town Gas followed in 1842, as gasification met demand for lighting in a pre-electric, pre-natural gas distribution system world. For the remainder of the 1800s and into the 1900s, additional gasification processes were developed. The Lurgi gasification process was developed in the 1880s. Gasification of coal for Fischer-Tropsch (F-T) liquids was used by Germany and Japan during World War II to produce fuels when oil supplies were limited. Production of Fischer-Tropsch fuels continued in the 1950s under South Africa's Sasol I, followed by other Sasol developments in the 1970s and 1980s. Over the last 50 years, there has been a steady progression of projects internationally (Figure 1). In each case, gasification-derived fuels have been developed to meet consumer needs, sometimes providing an alternative when oil-based fuel and natural gas supplies were constrained.

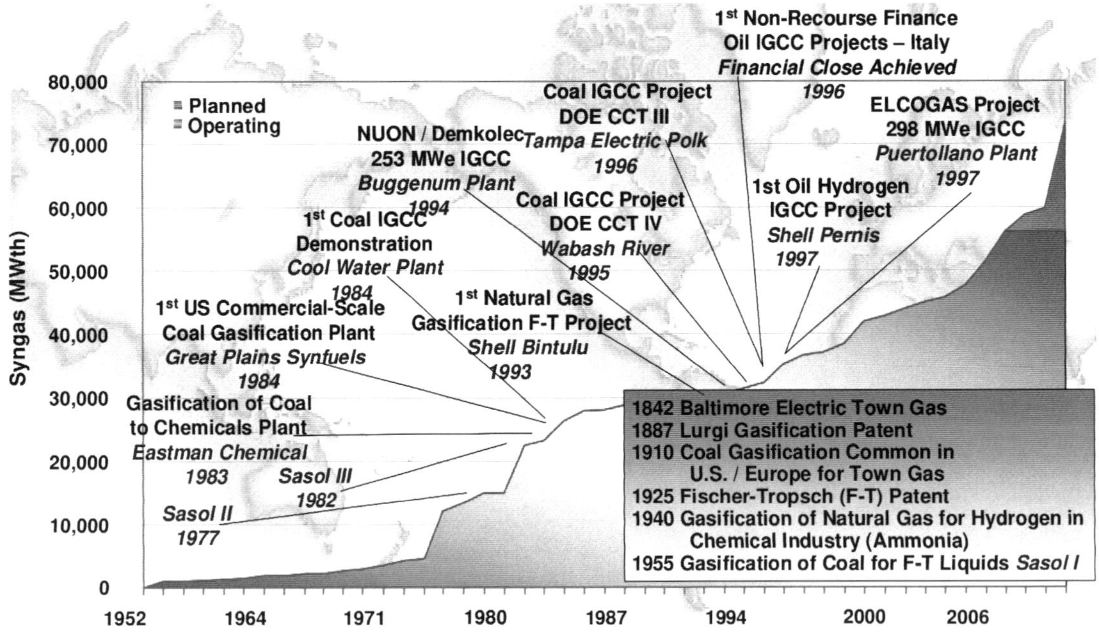

Figure 1: Progression of IGCC Projects over the Last 50 Years

2.0 DOE and IGCC - Pathway to Energy-Petrochemicals Security-Sustainability

The foundation of the U.S. Department of Energy (DOE) fuels policy rests on four cornerstones: (1) energy diversity to reduce dependence on oil; (2) reducing environmental impact and greenhouse gas (GHG) emissions; (3) enhancing the flexibility and reliability of the U.S. energy supply infrastructure to meet energy demand; and (4) strategies for cost-effectively improving the energy efficiency of the U.S. economy, thus reducing fuel demand. In pursuit of this policy, DOE has sponsored research and development (R&D) of a portfolio of technology solutions. Since the early 1980s, DOE has placed increasing emphasis on development of gasification technology as an energy security and sustainable fuel solution.

As high prices for natural gas and highway transportation fuels persist, coal gasification has offers an increasing attractive option to convert our nation's abundant coal resources into synthetic natural gas, transportation fuels, and hydrogen for fuel cells. Also, DOE has fostered development of coal-based integrated gasification combined cycle (IGCC) technology, which combines gasification with combined cycle generation of electricity. IGCC is energy efficient and has excellent environmental performance, and has been demonstrated, with support from DOE's Clean Coal Technology Program, and proven out in two commercial-scale plants operated by U.S. utilities.

FutureGen is DOE's vision for zero emissions power production, including carbon capture and storage (CCS) technology. Working with industry, the U.S. government plans to use federal funding to demonstrate cutting-edge CCS technology in commercial-scale IGCC and advanced coal power plants. The intent of the initiative is to build multiple commercial-scale generating plants, each with electricity generation capacity of at least 300 megawatts. DOE released a Funding Opportunity Announcement (FOA) valued up to $1.3 billion on June 24, 2008, for investment in IGCC/CCS generating plants. The FOA timeline includes return of proposals to DOE by October 8, 2008, and anticipated project selection by December 2008.

2.1 DOE Gasification Program and IGCC

DOE's gasification technology research program addresses the challenges of developing low-cost systems with zero emissions and high energy efficiency. The effort to reduce pollutants in the IGCC plants includes nitrogen oxides (NOx), sulfur oxides (SOx), arsenic (As), cadmium (Cd), selenium (Se), particulates, and carbon dioxide (CO_2). DOE has defined R&D pathways to achieve these goals. For instance, for IGCC plants coming online in 2010, goals include achieving 85% capacity factor and 98% carbon conversion. By 2015, additional technical advances (e.g., chemical looping gasifiers that concentrate H_2 and CO_2 streams without costly and energy intensive cryogenic oxygen production, hydrogen gas turbines, and solid oxide fuel cell topping cycle) would enable a 90% capacity factor. For IGCC systems coming online by 2010 and 2015 (2015 plants without carbon capture), DOE's performance and capital cost targets are:

- Energy Efficiency – Increase from 45-50% (HHV) by 2010 to 50-60% (HHV) by 2015
- Capital Cost – Decrease from $1000/kW by 2010 to $900/kW by 2015 (year 2002 dollars)

Additionally, by 2012, DOE's target is to demonstrate IGCC systems with CO_2 capture with no more than a 10% increase in the cost of electricity.

To reach its targets, DOE's R&D programs are addressing major technical issues involving gas cleaning, oxygen separation membranes, and hydrogen-carbon dioxide separation (as a post-water-gas shift reaction process) technologies and the use of low-rank coal (Figure 2).

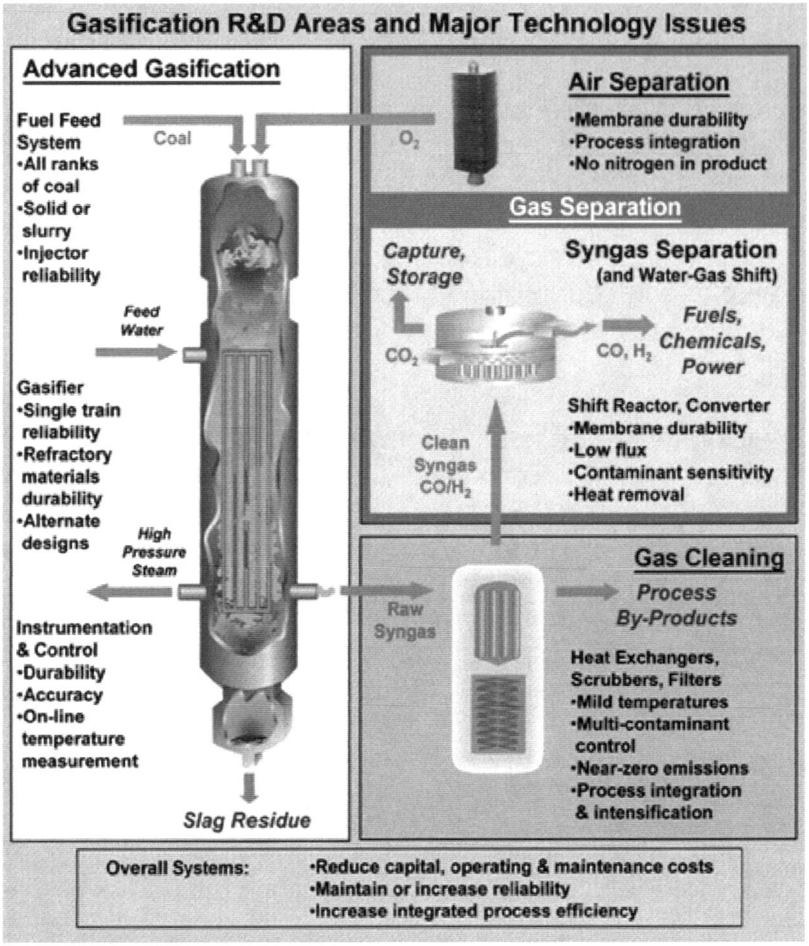

Figure 2: Gasification R&D Areas and Major Technology Issues

2.2 Flexibility of Gasification Feedstock

The flexibility of the gasification process opens the door to using energy feedstocks other than coal. Gasification processes can use sustainable renewable biomass, petroleum refining byproducts (e.g., pet coke and heavy reside), and waste products (e.g., refuse-derived fuels and municipal solid waste) as feedstocks, thereby converting these low- or negative-value feedstocks into high-value marketable fuels and chemicals.

2.3 Global Warming Considerations – Managing the Carbon Footprint: Carbon Sequestration

Based on detailed engineering studies by DOE's National Energy Technology Laboratory (NETL) and many other organizations, IGCC plants are likely the cheapest coal option when the power system cost equation includes CCS. For example, DOE has estimated that the addition of carbon capture to the average IGCC plant will increase the cost of electricity generation by 36%, but this increase is offset by the $42 per tonne CO_2 avoided cost.

2.4 From Gasification to Petrochemicals

Output of the gasification process includes a variety of chemicals that are building blocks for a multitude of consumer products (Figure 3). Syngas, a carbon monoxide and hydrogen mixture, that is the primary gasification process output, is a source of hydrogen and can be used in F-T syntheses to produce other chemicals, including naphtha, high-cetane diesel, jet fuel, and waxes. Conventionally, naphtha, a feedstock for producing basic petrochemicals such as ethylene and propylene, is a product of petroleum refining. Thus, the gasification/F-T path to naphtha promotes petroleum-independent production diversity. Examples of other gasification process by-product applications include:

- Slag for construction materials
- Hydrogen sulfide, leading to sulfur and sulfuric acid production
- Nitrogen for ammonia-based fertilizers
- Syngas conversion to synthetic natural gas
- Syngas conversion to methanol and ethanol

In each case, these applications offer potential for feedstock diversification away from fossil fuel alternatives.

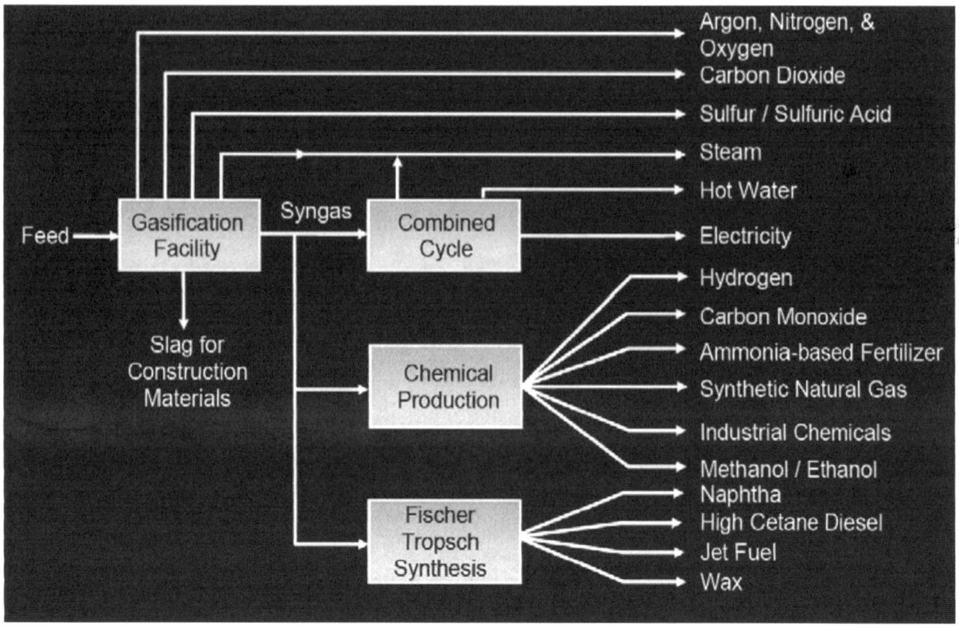

Figure 3: Gasification as a Source of Chemical Products

3.0 Progress Leveraged by International Collaboration

DOE takes an active role in fostering international cooperative efforts. Examples of organizations in which it takes a leadership role include the following:

Carbon Sequestration Leadership Forum (CSLF) – CSLF http://www.cslforum.org/ is an international climate change initiative for developing cost-effective carbon sequestration technology. It also addresses the regulatory environment needed to develop and commercialize such technology. CSLF includes 21 member countries and the European

Commission. The CSLF fosters R&D cooperation among member countries and has Task Forces that address Capacity Building in Developing Economies and Financing. The Capacity Building Task Force has conducted workshops in Brazil (2), Saudi Arabia, and the United States.

Asia Pacific Partnership on Clean Development and Climate (APP) – Government, industry, and academia in the seven APP **www.asiapacificpartnership.org** member countries – Australia, Canada, China, India, Japan, Korea, and the United States – are cooperating to address growing energy needs and related issues, such as energy security, environment protection, and climate change. The Partnership has established public-private Task Forces in eight key sectors: (1) cleaner fossil energy; (2) renewable energy and distributed generation; (3) power generation and transmission; (4) steel; (5) aluminum; (6) cement; (7) coal mining; and (8) buildings and appliances. The Task Forces are designed to meet Partnership goals through international cooperation to facilitate the development, diffusion, deployment, and transfer of existing, emerging and longer term cost-effective, cleaner, more efficient technologies and practices among the Partners through concrete and substantial cooperation so as to achieve practical results. Advanced power generation technologies, including supercritical and ultra-supercritical pulverized coal, IGCC, and oxy-fuel combustion, along with CCS have been identified by the Cleaner Fossil Energy (CFE) Task Force as priority areas for cooperation. Under this Task Force, several cooperative IGCC and CCS R&D and capacity building projects have been endorsed. Several joint R&D projects have been initiated between organization's member countries. Also, several workshops and seminars have been held:

- Integrated Gasification Combined Cycle with Carbon Capture and Storage (IGCC/CCS) Workshop and Design Information for APP Country Coals (Tokyo, Japan, September 4, 2007)
- Seminar on Integrated Gasification Combined Cycle (IGCC) combined with Carbon Capture and Storage (Melbourne, Australia, March 31, 2008)
- Workshop on Ultra-Supercritical Pulverized Coal and Carbon Capture and Storage (Melbourne, Australia, March 31, 2008)
- Seminar on Carbon Storage Technologies (Melbourne, Australia, April 1, 2008)
- Seminar on Technologies to Improve the Environmental Performance of Power Generation (April 1, 2008)

Asia-Pacific Economic Cooperation (APEC) – The United States has been a guiding force since the founding of APEC (http://www.apec.org/). In 2007, APEC Member Economies issued a Declaration on Climate Change, Energy Security and Clean Development. The declaration included a program of practical, cooperative actions and initiatives. The Expert Group on Clean Fossil Energy, which has been chaired by the DOE Office of Fossil Energy since its establishment in the early 1990s, is conducting a vibrant program of technical and policy studies, seminar, and capacity building activities. The EGCFE (www.egcfe.ewg.apec.org) completed the first-ever analysis of the carbon storage prospectivity of Southeast Asia and conducted capacity building workshops in Korea, China, and Mexico. Two additional workshops are being planned.

International Partnership for the Hydrogen Economy (IPHE) – The IPHE www.iphe.org was founded in 2003 to accelerate transition to an international hydrogen economy. It currently includes 16 member nations from around the globe and the European Commission. IPHE enables partners to coordinate hydrogen and fuel cell and provides a forum for members to address common codes and standards. The United States, through DOE, is a member.

International Energy Agency (IEA) – The IEA (www.iea.org) advises its 27 member countries and the European Commission on energy policy. In addition to developing energy market statistics, IEA conducts a broad fossil and renewable energy research program involving member country participants. It also is involved in energy technology collaboration and outreach to non-member countries such as China, India, Russia, and the OPEC countries. As a member country, the United States, through DOE, participates in many of the IEA's research activities.

Under the IEA's Working Party on Fossil Fuels (WPFF, www.iea.org/about/wpff.asp), the Task Force on Zero Emissions Technologies Strategy (ZETS) was begun in 2001 as a key strategic activity to accelerate the development and deployment of these technologies (e.g., IGCC and CCS). The ZETS initiative has been characterized by regular planning, workshops, and reports to address science and technology challenges and topics and implementation strategies. DOE plays a major role in ZETS through its chairing the WPFF.

Another area where the Office of Fossil Energy cooperates with the IEA is the Greenhouse Gas Programme (http://www.ieagreen.org.uk/), which was established in 1991. The Programme's three main activities are:

- Evaluation of technologies aimed at reducing greenhouse gas emissions,
- Promotion and dissemination of results and data from its evaluation studies, and
- Facilitating practical RD&D.

The IEA GHG Programme's activities to date have covered all the main anthropogenic greenhouse gases (CO_2, CH_4, nitrous oxide, and high global warming potential gases). Its current work is focused on ways to control and reduce CO_2 emissions from fossil energy facilities.

DOE is a major supporter of the IEA's Clean Coal Centre (www.iea-coal.org.uk). A collaborative project established in 1975, this Centre supports its member's efforts to make the production, transportation, and use of coal sustainable. It accomplishes this through a program of work that includes studies, reports, newsletters, seminars and workshops that are of considerable significance for all countries involved in the use or supply of coal. Among its notable activities is the Centre's annual International Conference on Clean Coal Technologies for our Future.

In addition, DOE's National Energy Technology Laboratory has recently signed bilateral agreements with several key countries on advanced coal conversion and CCS, including:

- Ministry of Science & Technology (China)
- Pontifical Catholic University of Rio Grande do Sul (Brazil)
- Brazilian Coal Association (Brazil)
- Interdisciplinary Center, Herzliya (Israel)

- Central Mining Institute/Institute for Chemical Processing of Coal (Poland)

Several additional agreements with China and India are being developed.

A century of petrochemical advances

Peter R Pujadó, consultant, Kildeer, IL
Bipin V Vora, consultant, Naperville, IL

Abstract

The beginnings of the petrochemical industry date back more than 150 years. The first chemicals produced were aniline and dyestuffs and other derivatives from coal starting in the mid-1800s. Then, scientists in Germany led major advances, starting from acetylene production via Reppe's chemistry. This presentation gives an overview from these early beginnings to modern day cracking for ethylene and propylene production and modern aromatics complexes, derivatives, and the like, to next generation technologies on the horizon, such as those based on natural gas or on coal to olefins via syn gas and methanol intermediates.

Historical perspective

We think it is fair to claim that the petrochemical industry was born and developed as a result of the dyestuff industry, or rather from the quest for synthetic dyes, in particular synthetic indigo, a naturally occurring blue dye extracted from a number of plants in SE Asia, and in Central and South America. While natural indigo was used throughout the XIX century, today all indigo-type dyes are synthetic.

The main precursor of all modern synthetic dyes was aniline, the simplest aromatic amine. The following was liberally abstracted from a Wikipedia contribution on “aniline”:

> Aniline was first isolated from the destructive distillation of indigo in 1826 by Otto Unverdorben, who named it *crystalline*. In 1834, Friedrich Runge (Pogg. Ann., 1834, 31, p. 65; 32, p. 331) isolated from coal tar a substance that produced a beautiful blue color on treatment with chloride of lime, which he named *kyanol* or *cyanol*. In 1841, C. J. Fritzsche showed that, by treating indigo with caustic potash, it yielded an oil, which he named *aniline*, from the specific name of one of the indigo-yielding plants, *Indigofera anil*, anil being derived from the Sanskrit nīla, dark-blue, and nīlā, the indigo plant. About the same time N. N. Zinin found that, on reducing nitrobenzene, a base was formed, which he named *benzidam*. August Wilhelm von Hoffmann investigated these variously-prepared substances, and proved them to be identical (1855), and thenceforth they took their place as one body, under the name *aniline* or *phenylamine*.
>
> Its first industrial-scale use was in the manufacture of *mauveine*, a purple dye discovered in 1856 by Hoffmann's student William Henry Perkin (for whom the Perkin medal was later named by the Society of Chemical Industry to honor outstanding contributions in the field of industrial chemistry). The great German chemist von Hoffmann (whose name is variously spelt Hoffman or Hofmann) had been brought over to England by queen Victoria’s prince consort, Albert, to teach at the Royal College of Chemistry in London. At the time of *mauveine*'s discovery, aniline was an expensive laboratory compound, but it was soon prepared "by the ton" using a process

previously discovered by Antoine Béchamp. The synthetic dye industry grew rapidly as new aniline-based dyes were discovered in the late 1850s and 1860s.

On this basis, we can place the birth of the petrochemical industry in the mid-1800s. However, a parallel chemical or "petrochemical" industry had derived in parallel in the meantime based on the recovery of fractions from coal tar distillation, principally in Germany and around Manchester, in Lancashire. Before we proceed we must give credit to Peter H Spitz for his volume "Petrochemicals – The rise of an Industry," published by John Wiley and Sons in 1988. From here onwards, any quotations or references derived from this book will be simply denoted by (*).

Coal tar and coke oven gases had been used to produce chemicals in Lancashire as early as 1817 (*), A number of compounds were separated and identified. In addition to the basic aromatics, from benzene to naphthalene and anthracene, other derivatives were produced: nitrobenzene, phenol, picric acid, and others. Subsequent efforts, however, centered on the synthesis of dyestuffs, and in this area, the German chemical industry soon took an undisputed leadership. After beating Perkin's patent application for the synthesis of alizarin by one day (*) the German chemists proceeded to commercialize symtheses for methylene blue, scarlet red, benzopurpurin, etc. Not surprisingly, the three larger German chemical firms that formed the huge I.G.Farbenindustrie chemical combine all had names associated with dyestruffs : Badische **Anilin** und Soda Fabrik, **Farb**werke Hoechst, and **Farben**fabriken Bayer, as did the combine itself (I.G. **Farben**).

In reality, then, the petrochemical industry derived from coal and is thus aptly named. It evolved along a number of semi-independent branches, not all simultaneously. Most of the early processes were developed by industrial chemists ("chemical engineering" as such did not yet exist) in Germany and were based on the use of acetylene and benzene (and the other aromatics). Although known, ethylene at the time was far from being an industrial commodity.

Over and over again we come to realize that the petrochemical industry largely developed in Germany and was based on coal. The petrochemical industry that developed in the Unite States in the 1930s and 1940s had the advantage of having virtually unlimited supplies of petroleum and natural gas in the US Gulf Coast and thus gave the US a significant edge in the war effort over Germany. Although a lot was known about the prowess of the German petrochemical industry, it was only at the end of World War II that its full vastness, inventiveness, and creativity were revealed when the CIOS (Combined Intelligence Objectives Subcommittee) teams carefully described the products and technologies at the many German plants, principally at I.G. Farben (BASF, Hoechst, Bayer, and others), Chemische Werke Hüls, etc

The first large-scale synthetic petrochemical was acetylene. Production of acetylene by the hydrolysis of calcium carbide had been shown by a French scientist, Moissan, in 1892. Calcium carbide was produced by the high-temperature processing of coal and limestone mixtures in an electric furnace. This technology was widely used in Germany and other countries (e.g. Union Carbide in the United States); its current use is pretty much obsolete and discontinued, but at least only a few years ago (in the early 1990s) a plant was still operational in Guizhou, China.

Table 1

Historical dates of the first commercial production of some important organic chemicals (*)

phenol	F Raschig (Germany)	1901
	Hoffmann-LaRoche (Switzerland)	
carbon tetrachloride	Griesheim-Elektro (Germany)	1903
trichloroethylene	Wacker (Germany)	1908
ethylene	Griesheim-Elektro (Germany	1913
ammonia	BASF (Germany)	1913
acetic acid	Wacker (Germany)	1916
ethylene oxide	BASF (Germany)	1916
acetaldehyde	Hoechst (Germany)	1916
acetone	Hoechst (Germany)	1917
	Weitzman (UK)	
	Standard Oil of NJ (US)	
vinyl acetate	Shawinigan Chemicals (Canada)	1920
methanol	BASF (Germany)	1923
butanol	BASF (Germany)	1923
vinyl chloride	Wacker (Germany)	1930

As early as 1863 another French chemist, Berthelot, had discovered that acetylene can be produced by passing methane gas across an electric arc. This technology was finally incorporated into the industrial production of acetylene in Germany in the mid-1920s and early 1930s, and, in a more modern version, is the technology used by Chemische Werke Hüls and its successor companies (Oxeno, etc.) in Marl, Germany. In the United States a similar technology was commercialized By the Standard Oil Company (New Jersey) (now Exxon-Mobil) in co-operation with I.G. Farben.

Acetylene was primarily used to produce acetaldehyde using a mercury-sulfuric acid catalyst (still in use at Guizhou) that could be converted to acetic acid. More importantly, however, was the alkaline aldol condensation of acetaldehyde followed by hydrogenation to butylenes glycol over a copper-chrome catalyst (*). Dehydration to butadiene was effected over a phosphate catalyst supported on coke. Butadiene is the main component in the production of synthetic rubber of vital importance in the war effort.

Ethylene could be recovered as a byproduct from the production of acetylene or by the selective hydrogenation of ethylene. Ethylene was alkylated with purified coke-oven benzene in the liquid phase with $AlCl_3$ catalyst (*) and the resulting ethylbenzene dehydrogenated to styrene over zinc oxide.

Interestingly, olefins were also produced in Germany by the Fischer Tropsch technology from CO/H_2 mixtures (synthesis gas or syn gas) obtained from a Winkler coal gasifier (*). These olefins ranged from ethylene to the C_{11} to C_{17} fraction and were use for the production of alcohols (and detergents) in what was probably the first Oxo unit ever built. This work was based on the discoveries of Franz Fischer and Hans Tropsch who in 1925 had demonstrated that syn gas could be converted into mixtures of liquid oxygenates and hydrocarbons at relatively mild operating conditions and, even today, is the best solution for a petrochemical and fuels industry that can become relatively independent from petroleum consumption. The first commercial Fischer-Tropsch unit was built by Ruhrchemie in Holten in 1934 (*).

Although not a petrochemical as such, ammonia plays a critical role in the petrochemical industry, Here too the contributions of German industrial chemists were fundamental. The ideas developed by Fritz Haber and the reactor design by Carl Bosch resulted in the commercialization of the so-called Haber-Bosch ammonia synthesis process that first came on stream in 1913. Perhaps lesser known are the catalyst discoveries contributed by Alwin Mittasch (*) (who had first synthesized methanol from syn gas as early as 1913) who supposedly tested over 20,000 different catalytic materials (an early version of combinatorial chemistry!) to finally arrive at a catalyst composed of magnetite with a few percent aluminum and a small amount of potassium.

In all of the above, German industrial chemists made great strides at understanding and developing the role of catalysts in chemical syntheses. This also contributed to the development of technologies for the production of a large variety of chemicals: phthalic anhydride by the liquid-phase oxidation of naphthalene with a mercury catalyst (only later replaced by the vapor-phase oxidation over vanadium pentoxide); sulfuric acid by the direct oxidation of sulfur dioxide with air; production of hydrocarbon fuels by the direct hydrogenation of coal; the synthesis of methanol and higher alcohols from synthesis gas using a zinc-chromium catalyst; high-pressure urea synthesis; various applications of carbonylation and hydroformylation reactions; hydroforming of paraffins to aromatics (a precursor of later catalytic reforming technologies), etc. First and foremost was perhaps the development of the so-called "Reppe chemistry" after Walter Julius Reppe (1892-1969) that exploited a number of interesting applications based on acetylene. The following is copied from Wikipedia:

The high pressure reactions catalysed by heavy metal acetylides, especially copper acetylide, or metal carbonyls is called Reppe Chemistry. Reactions can be classified into four large classes:

- Vinylization according to the equation:

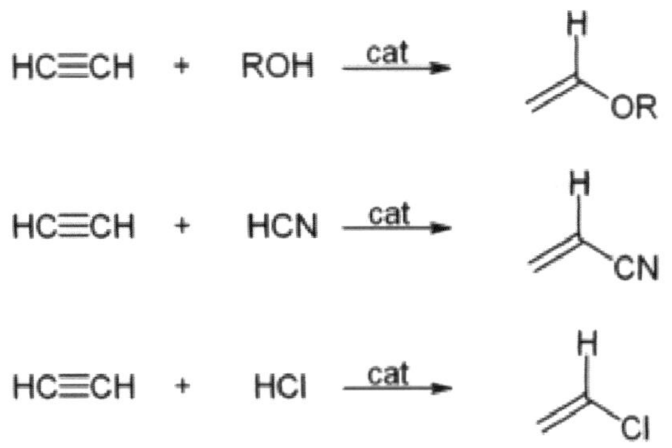

- Preparing ethynyldiols from aldehydes according to the equation:

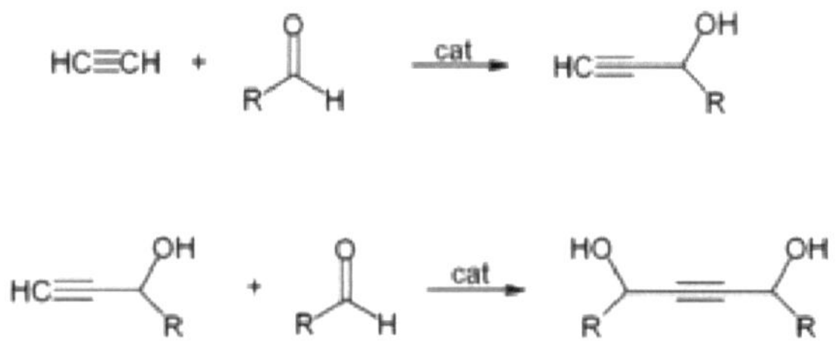

- Reactions with carbon monoxide:

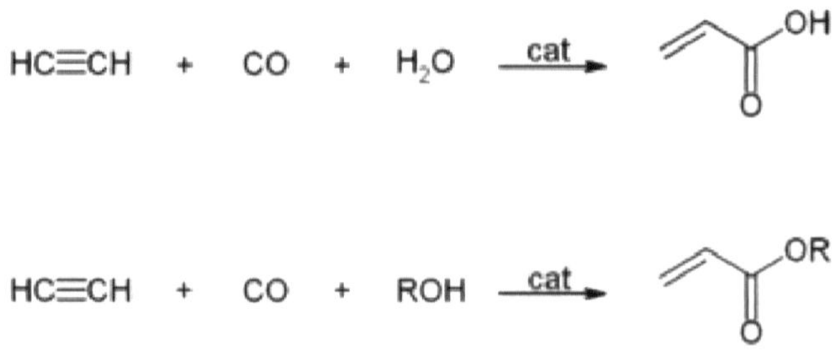

This simple synthesis was used to prepare acrylic acid derivatives for the production of acrylic glass.

- Cyclic polymerization of acetylene

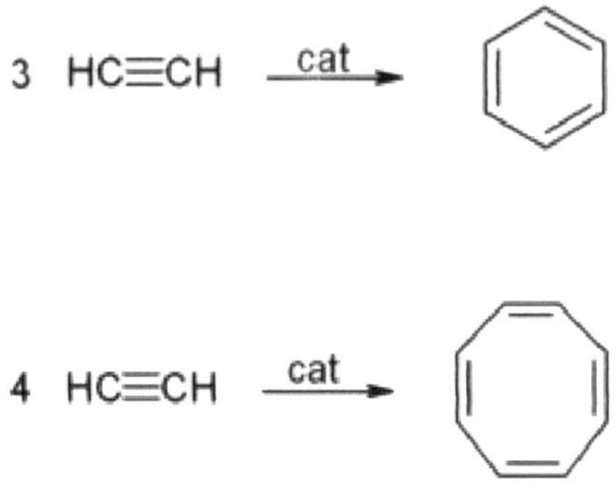

This reaction provided an unusual route to benzene and especially to cyclooctatetraene, which was difficult to prepare otherwise.

Products from these four reaction types proved to be versatile intermediates in the syntheses of lacquers, adhesives, foam materials, textile fibers, and pharmaceuticals that could now be produced.

Not to be overlooked also were the early developments of plastics and synthetic fibers: nylon 6,6 from adipic acid and hexamethylenediamine (HMDA) developed by Wallace Hume Carothers at DuPont (1935), or nylon 6 from the self-polymerization of caprolactam developed by Paul Schlack at I.G. Farben (1938), or the early polyesters (also by Carothers although initially dismissed by him as impractical), polyurethanes (initially from hexamethylene-di-isocyanate and 1,4-butanediol), and many others ranging from poly-vinyl chloride to chloroprene or neoprene.

Today's petrochemical industry

Today's petrochemical industry is perhaps better viewed as families of compounds derived from a starting feedstock. The main starting feedstocks are olefins and aromatics. An alternative starting feedstock is synthesis gas or "syn gas" (mixtures of H_2 and CO) that can be obtained from any hydrocarbonaceous material ranging from methane to coal.

While the early petrochemical industry was based on acetylene, modern petrochemicals are derived from olefins. Although still very reactive, olefins are much less energetic than acetylene and, thus, more stable, safer, and economic to produce. Invariably, most olefins are produced by the cracking of hydrocarbon feedstocks, whether thermal or catalytic. An alternative route, still not fully commercialized, produces olefins by the catalytic conversion of methanol (from syn gas). A relatively obsolete route (for economic reasons) is the production of ethylene by the dehydration of ethanol from bio sources.

Thermal cracking of hydrocarbons in the presence of dilution steam (steam cracking)is by far the largest source of olefins. If the hydrocarbon is ethane, the product olefin is ethylene obtained in high yield. If the feed hydrocarbon is heavier – propane, butane, naphthas, or even gas oil – the product olefins range from ethylene to C_5s and aromatics. Typically, the yield of ethylene decreases with increasingly heavier feedstocks, while the yield of propylene remains fairly uniform. Because of the variety of valuable coproducts and byproducts obtained from steam cracking, this type of unit invariable is the focal point of very large industrial complexes. While the first steam cracking units built in the early 1940s had relatively small ethylene capacities (6000 to 10000 t/year), modern large-scale complexes routinely produce in excess of 500,000 t/year ethylene. The major coproducts and byproducts include:

- propylene
- 1,3-butadiene
- Isobutylene
- 1-butene (and perhaps 2-butene)
- py gas – a C_5+ cut rich in olefins and aromatics from which various olefins and diolefins (e.g., isoprene) can be recovered, as well as BTX (benzene, toluene, and xylenes) and heavier aromatics

Ethylene and propylene (and, if recovered, 1-butene) are routinely upgraded to polymer-grade quality (at least 99.95% purity in the case of ethylene) by means of fractionation . The products are usually shipped to separate polymerization facilities but such facilities are sometimes integrated within the complex. All the products from s steam cracker find use in a myriad of petrochemical derivatives. To name but a few:

- from ethylene
    - ethylene oxide – by direct catalytic oxidation
    - ethylene glycol – by the hydration of ethylene oxide
    - ethylene dichloride/vinyl chloride – via chlorination and oxychlorination
    - acetaldehyde – by ethylene oxidation in either a one-step or a two-step process
- from propylene
    - propylene oxide – by propylene epoxidation, either via the chlorohydrins process or by means of a hydroperoxide (isobutane hydroperoxide to yield *tert*-butanol or MTBE as a coproduct, or, more importantly, ehylbenzene hydroperoxide to yield styrene as a coproduct)
    - propylene glycol – via propylene oxide hydration
    - acrylonitrile – via the ammoxidation of propylene; acetonitrile and hydrogen cyanide are major valuable coproducts; acrylonitrile can also be dimerized to adiponitrile via an electrochemical or a chemical process
    - acrolein and acrylic acid – by direct catalytic oxidation; methionine is also an important derivative of acrolein
- from butenes
    - *tert*-butanol or MTBE from isobutylene; methacrylayes can be prepared from *t*-butanol
    - *sec*-butanol and methyl ethyl ketone (MEK), usually from 2-butene
    - maleic anhydride, currently mostly obtained by the catalytic oxidation of n-butane and an intermediate in the production of 1,4-butanediol
    - high-octane motor fuel alkylate by the alkylation of 2-butene with isobutane
    - high-octane motor fuel gasoline by the condensation of isobutylene with itself or with other butenes, followed by hydrogenation of the resulting products
    - 1,3-butadiene for the production of various synthetic rubbers and polymers (polybutadiene, acrylonitrile-butadiene-styrene, etc.)
    - adiponitrile from 1,3-butadieneby hydrocyanation; adiponitrile can be used to produce hexamethylene diamine (HMDA)
- from benzene
    - ethylbenzene and styrene
    - cumene (isopropyl benzene) phenol and acetone (acetophenone is a byproduct)
        - from acetone – methacrolein, methacrylic acid, and methacrylates
        - from acetone – ketene and acetic anhydride
        - from acetone and phenol – bisphenol A and polycarbonates
        - from phenol –aniline (alternative route)
        - from phenol – cyclohexanol and cyclohexanone (alternative route)
    - cyclohexane
        - cyclohexanol and cyclohexanone, intermediates for the production of caprolactam (and nylon 6)

        - adipic acid (also from phenol and other sources); my be used in theproduction of adiponitrile, 1,6-hexanediol, 6-hydroxycaproic acid, etc.
    - nitrobenzene and aniline (also from phenol); also diamines; anilne and formaldehyde can be condensed to 4,4'-diaminodipheylmethanes which, once phosgenated, yield MDI (methylene diamino diphenyl isocyanateor diphenyl methane 4,4'-diisocianate)
    - maleic anhydride – currently mostly produced by n-butane oxidation; intermediate in the production of 1,4-butanediol, tetrahydrofuran, *gamma*-butyro lactone, fumaric acid, etc.
- from toluene
    - benzoic acid
    - nitrotoluene, dinitrotoluene, and trinitrotoluene (TNT)
    - tolylene diamines and tolylene diisocyanate (TDI) for the production of polyurethanes
    - terephthalic acid via toluene carbonylation (alternative scheme)
- from o-xylene
    - phthalic anhydride and many plasticizers, alkyd resins, unsaturated polyester resins, etc.
    - terephthalic acid (alternative scheme)
- from p-xylene
    - terephthalic acid (TPA and PTA) (and dimethylterephthalate, DMT)

The above summary provides a bird's eye view of the major petrochemicals derived from steam cracking. Steam cracking, however, is not the only source of petrochemicals. Others are:

- fluid catalytic cracking (FCC) is major source of propylene
- catalytic reforming of naphthas is the main source of BTX aromatics
- hydrocyclodimerization of propane and butanes is an alternative source of BTX aromatics
- methanol derived from syn gas is the main source of acetic acid and derivatives

Syn gas (blends of hydrogen and carbon monoxide, CO) is a very versatile, though indirect, source of petrochemicals

- syn gas can be used to produce olefins and saturated hydrocarbons via the Fischer-Tropsch reaction
- syn gas is required for the synthesis of methanol and derivatives (e.g., formaldehyde)
- methanol from syn gas can be used for the production of olefins (mostly ethylene and propylene) and aromatics (gasoline and heavy aromatics like durene)
- hydrogen from syn gas is absolutely necessary for all hydrogenation reactions, desulfurization reactions, etc.
- carbon monoxide, CO, from syn gas is required for carbonylation reactions: carbonylation of methanol to acetic acid, carbonylation of toluene to p-toluic aldehyde (and terephthalic acid)
- carbon monoxide and hydrogen in the hydroformylation of olefins to aldehydes (and to alcohols or acids)
- carbon monoxide for the production of phosgene for the synthesis of isocyanates

No matter how comprehensive we try to make this list, there are always other compounds that are hard to fit in a logical sequence of feedstocks and products. For example,

- polyether polyols – from the reaction of an epoxide with a polyol (a glycol or glycerin, for example)
- 1,3-propanediol – a novel polymer intermediate to compete with ethylene glycol or 1,4-butanediol in the production of polyesters
- various specialty isocyanates – hydrogenated diphenylmethane 4,4'-diidocyanate (HMDI), hexamethylene diisocyanate (HDI), xylylene diisocyanate (XDI), hydrogenated xylylene diisocyanate (HXDI), naphthalene diisocyanate (NDI), etc.
- 2,6-naphthalene dicarboxylic acid – a naphthalene-based analogue of terephthalic acid
- 1,4-dimethylol cyclohexane – a naphthenic glycol for polyester manufacture
- 1,4-butanediol by the Reppe reaction between acetylene and formaldehyde to 1,4-diol-2-butyne; 1,4-butanediol by the acetoxylation of 1,3-butadiene
- laurolactam for the manufacture of nylon-12 (usually from cyclododecane by butadiene cyclotrimerization)

Although this article concerns "petrochemicals" we should not forget that other non-petrochemical compounds are often required in the petrochemical industry. The two most important ones are perhaps

- ammonia – urea, nitric acid, NPK fertilizers, etc.
- chlorine – caustic an related compounds
- sulfuric acid and derivatives
- hydrogen peroxide – as used in several epoxidation reactions

A discussion of these separate industries is beyond the scope of this article.

References

1. Alain Chauvel and Gilles Lefebre, Petrochemical Processes, Vol 1 – Synthsis gas derivatives and major hydrocarbons, Vol 2 – Major oxygenated, chlorinated, and nitrated derivatives, Éditions Technip, 1985
2. G Margaret Wells, Handbook of Petrochemicals and Processes, 2nd edition, Ashgate Publishing Ltd, 1999

The Role of Industrial Gases in Fuels and Petrochemicals; Past, Present and Future

Dante P. Bonaquist, Senior Corporate Fellow and Chief Scientist, Praxair, Inc
dante_bonaquist@praxair.com

Introduction

Industrial gases (oxygen, H_2, CO, nitrogen and CO_2) have played an integral part in the development of the modern refining and petrochemical industries; yet they function quietly compared to more readily recognized feedstock such as crude oil, natural gas and ethylene. This is due in part to the nature of these gas molecules as fundamental utilitarian building blocks for chemical synthesis as well as their low profile roles as enhancers of safety, quality and productivity and in mitigation of environmental footprint.

Each of the major industrial gas supply modes including pipelines, dedicated on-site production units, refrigerated liquid transport and high pressure cylinders are involved in delivery of gases to refineries, chemical complexes and individual operations within those facilities. The supply modes cited above are capable of delivering cost effective quantities ranging from greater than 100 million scf/day in the case of H_2, as much as 10,000 tons/ day for oxygen or nitrogen to a few pounds per day for analytical gases and mixtures.

The goal of this paper is threefold:

Summarize primary industrial gas production methods.

Provide an overview of industrial gas utilization in refineries and petrochemical plants up to the present state by highlighting key applications.

Identify emerging technologies and applications that are made possible in part by reliable cost effective industrial gas supply.

Gas Production Methods

Oxygen and Nitrogen: For more than 100 years, the only means for producing significant quantities of oxygen and nitrogen for commercial applications has been air separation. Cryogenic distillation has grown with other measures of industrial production, both in total volume of air separated and scale of plants. Today, plants based on cryogenic distillation are capable of producing more than 3,000 tons per day of gaseous oxygen in a single train with or without by-products including gaseous nitrogen as well as liquid oxygen, nitrogen and argon. Oxygen purity in excess of 99.9% is common for plants supplying ethylene oxide facilities although 99.5 % has been the industry standard for general purpose oxygen plants, Special purpose plants have been developed to supply large volumes of gaseous nitrogen for enhanced oil recovery and oxygen for gasification processes. In the near future, large plants with single train capacity over 5,000 tons per day of oxygen are envisioned for huge clean coal combustion, gasification and coal to liquids facilities. Figure 1 is a schematic of a cryogenic air separation plant.

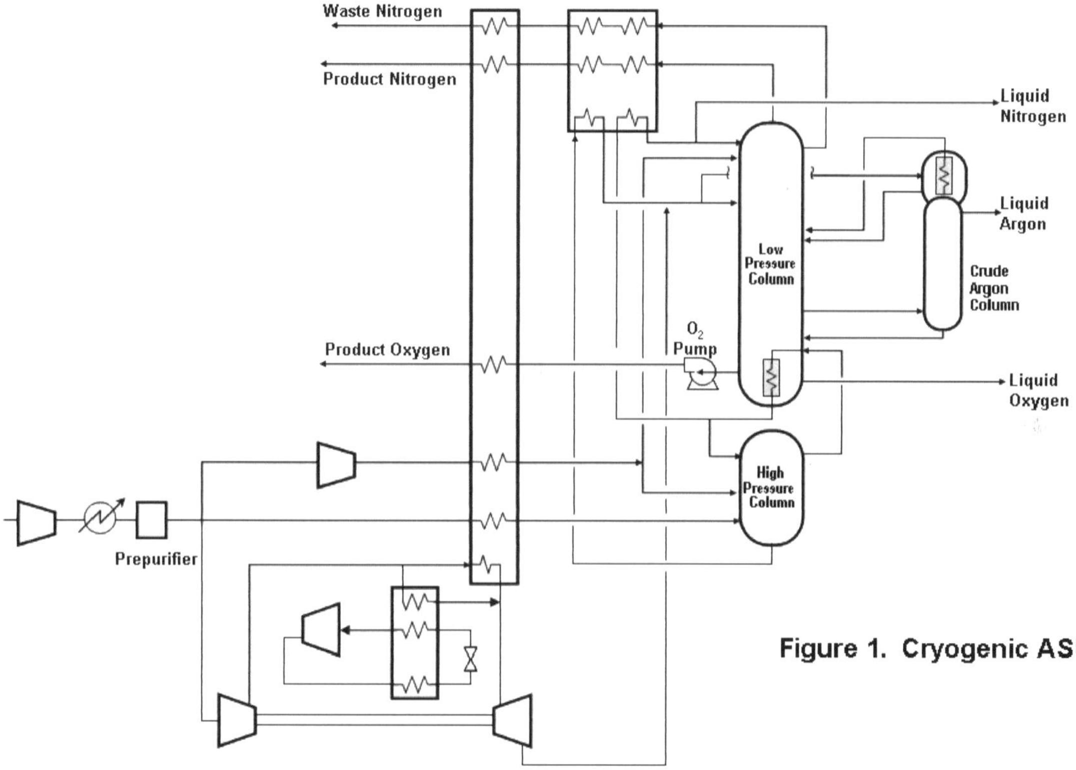

Figure 1. Cryogenic ASU

In the 1970's, pressure swing (PSA) and vacuum pressure swing adsorption (VPSA) were introduced for producing low purity oxygen (90%) for combustion applications. PSA is used to make low purity nitrogen (98%) for inert blanketing. The low specific energy and capital cost of the oxygen VPSA process make it the process of choice for supplying oxygen to applications that can accept lower purity. Initially introduced as a small scale oxygen generator, VPSAs have grown in capacity to 250 tons per day of oxygen for a single train. Figure 2 is a schematic of an oxygen VPSA unit.

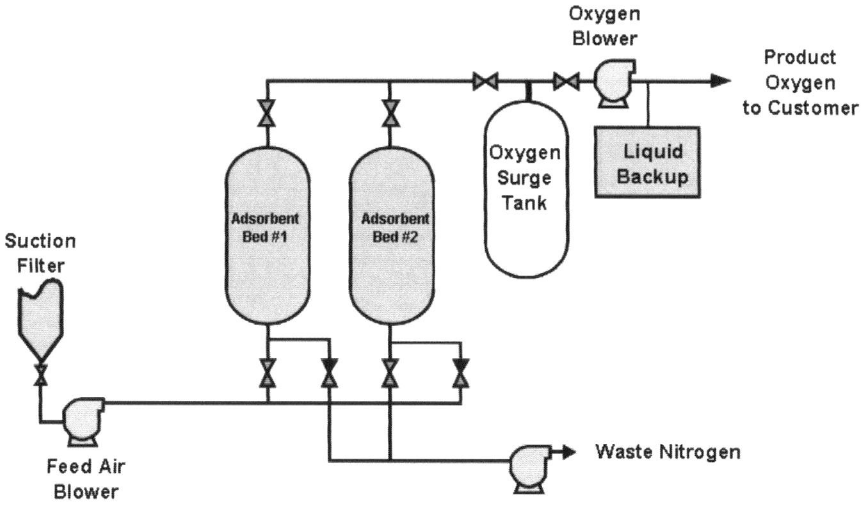

Figure 2. Two-bed VPSA

Commercial polymeric membrane based systems for supplying small to modest volumes of low purity nitrogen were introduced in the early 1980s by Dow (1). The size of the market served by membrane units has grown steadily. The primary attributes of this air separation technology include modular plant design and simplicity of operation. Applications are limited to those where levels of oxygen in the range of 0.5 to 5% can be tolerated.

Figures 3 and 4 show the commercial applicability of various air separation systems based on technologies discussed above.

Figure 3. Preferred Oxygen Supply System

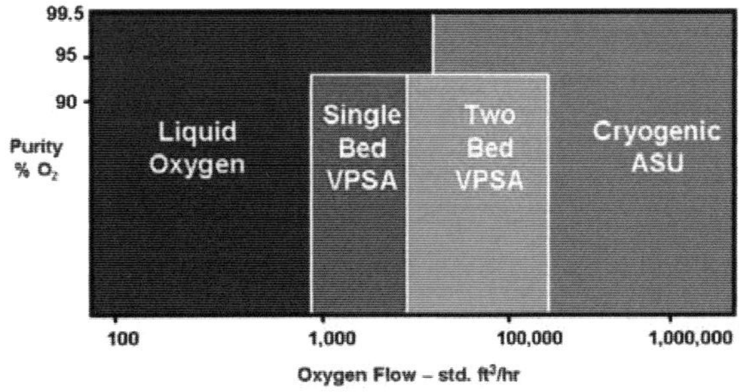

Figure 4. Preferred Nitrogen Supply System

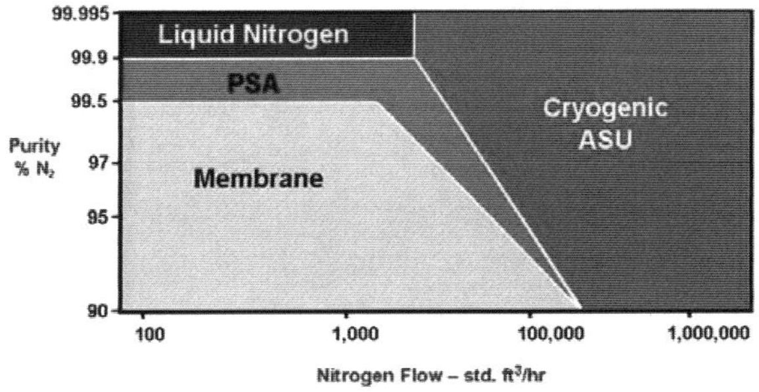

In addition to continuous improvement of current air separation technologies being pursued by all major industrial gas firms, Praxair and Air Products (2,3) are involved in research and development expected to lead to commercialization of air separation using high temperature oxygen conducting ceramic materials called Oxygen Transport Membranes OTM or Ion Transport Membranes (ITM). The primary focus for application of oxygen conducting ceramic membranes is gasification and combustion with CO_2 capture. Figure 5 illustrates the principle of ceramic membrane operation. In this example, the partial pressure driving force is setup by reacting oxygen with a fuel stream as it emerges from the membrane. Compression of air can also provide this driving force.

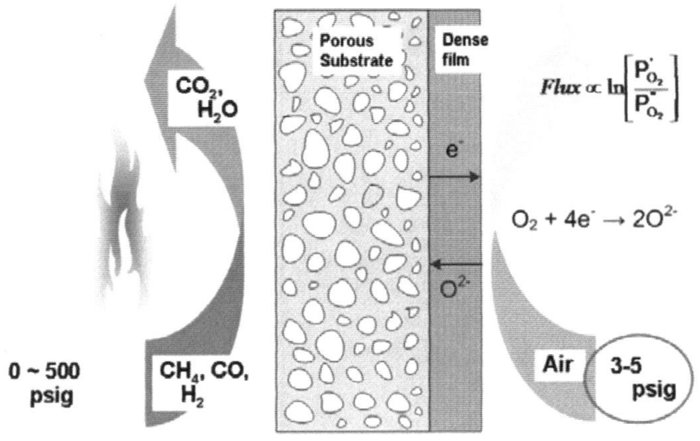

Figure 5. Oxygen Transport Membrane Used for Combustion

Hydrogen and CO and Syngas: H_2 and CO are related products made by reforming or partial oxidation of hydrocarbons. Other processes are used commercially to supply H_2, for example electrolysis of water; however, they are not relevant to refining and petrochemicals. H_2 and CO are commonly supplied together as synthesis gas for the production of methanol and oxo-alchools (4). The reactions involved for production of H_2 and CO by reforming and partial oxidation are:

Reforming: $CH_4 + H_2O$ + Heat ➔ 3 H_2 + CO (Catalyst: Ni, Ru, Rh, Pd, Pt)

Water Gas Shift: CO + H_2O ➔ CO_2 + H_2 + Heat (Catalyst: high T; Fe/Cr oxides, low T; Cu)

Partial Oxidation: $CH_4 + 0.5O_2$ ➔ CO + H_2 + Heat (Catalyst: Rh, Pt)

H_2 plants consist of a steam methane reformer (SMR) that converts the hydrocarbon carbon feedstock to H_2 and CO, followed by at least one stage of shift conversion to react most of the CO with steam to form CO_2 and more H_2. Plants design over 25 years ago generally use a second stage of shift conversion to increase yield. Purification of the H_2 product on the older plants is done by absorption of CO_2 using a solvent such as MEA followed by methanation of residual CO and CO_2. The product is about 97% pure.

In modern plants, the absorption and methanation steps are replaced by a multibed PSA for purification of the H_2 product stream to 99.99% or greater. The higher purity is generally beneficial to customers. The fuel requirements of the SMR furnace are met to a large degree by the PSA tailgas. To improve the thermally efficiency of the reforming process, steam is generally exported, often to the refinery that is the customer for H_2 (5). Figure 6 is a block flow diagram for a hydrogen plant.

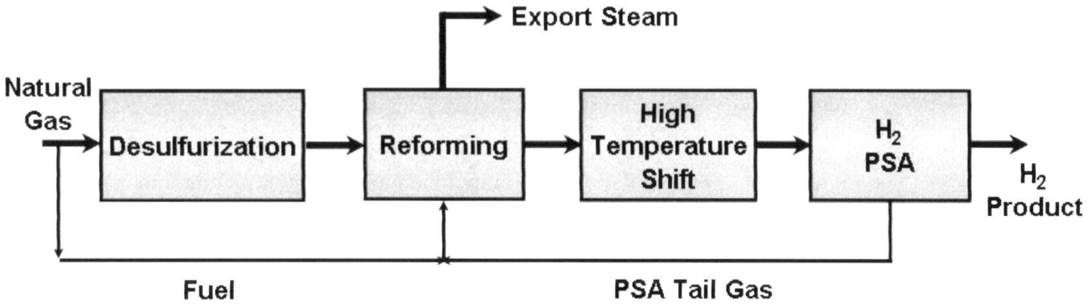

Figure 6. Hydrogen Plant Block Flow Diagram

Unlike oxygen and nitrogen that can be liquefied in large volumes to provide backup to air separation units, it is impractical to liquefy H_2 for this purpose. It is therefore advantageous for large H_2 users to receive at least part of their gas from a multi-plant pipeline network such as that operated by Praxair on the US Gulf Coast. As indicated in Figure 7, this system includes a first of a kind H_2 storage cavern created by solution mining of a salt formation that further increases system reliability (6).

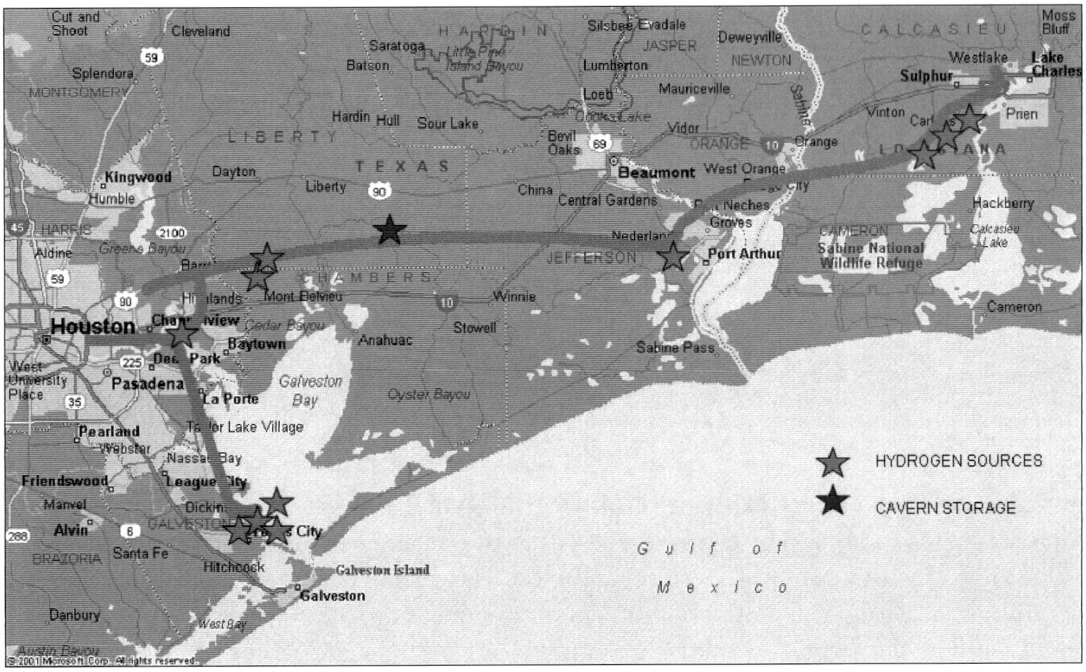

Figure 7. Praxair Gulf Coast Hydrogen System

In addition to "on purpose" production from SMR based plants, H_2 is recovered as a by-product from various refinery and chemical plant off-gases. This is generally economical using a PSA unit for streams containing 50% or more hydrogen by volume. Recently QuestAir working with ExxonMobil has introduced a specialized PSA unit for this purpose (7). A number of PSAs designed by UOP and others are in service for by-product H_2 recovery (8).

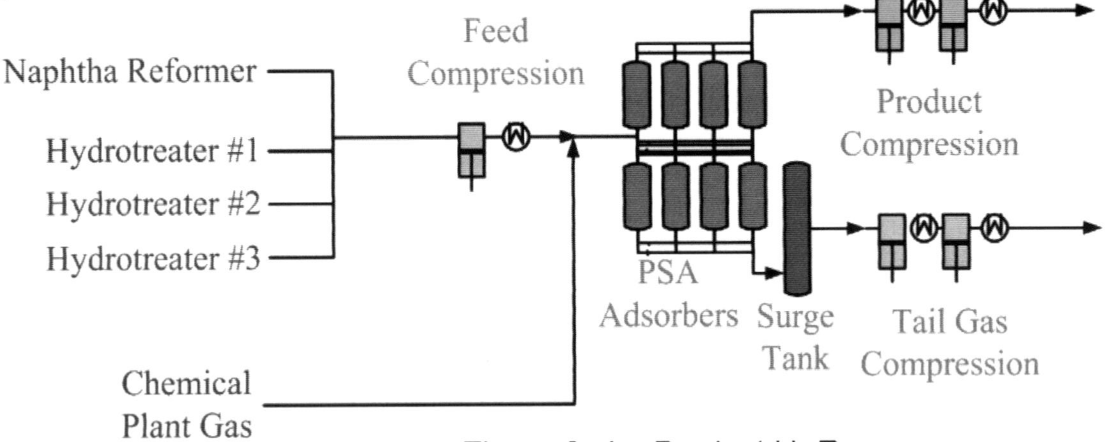

Figure 8. by-Product H_2 Recovery

Applications – Overview

I have selected several commercially significant refining and petrochemical process to illustrate the linkage between the development of configurations we see today and primary industrial gases. For a comprehensive treatment, you are encouraged to read Harold Gunardson's book, *Industrial Gases in Petrochemical Processing* (9).

Oxygen: Commercially important reactions involving oxygen, either as a pure component, or as oxygen enriched air include (10):

oxygen + paraffins, olefins ➔ aldehydes, alcohols
oxygen + olefins ➔ epoxides
oxygen + refining resid, petcoke, coal ➔ synthesis gas
oxygen + H_2 sulfide ➔ sulfur

Ethylene Oxide (EO): Ethylene for feed to EO processes is made by steam cracking of various hydrocarbons. In the US, ethane and propane are common starting points. The current selling price of ethylene (August 2008 spot) is $0.70/lb. or $1,400 per ton. To produce EO, ethylene is reacted with oxygen over a silver catalyst at 250 deg. C and 20 atmospheres of pressure.

Direct oxidation was first commercialized by Shell in the 1950s replacing the chlorohydrin process where ethylene is reacted with hypochlorous acid (chlorine and water) to make ethylene chlorohydrin. The oxide is formed by treatment of the chlorohydrin with slaked lime or caustic soda (11). Historically air has been used as a source of oxygen; however, state of the art EO plants take advantage of the reduced size and improved selectivity available with high purity oxygen. Higher purity oxygen, 99.95% compared to 99.5%, the industrial standard, will generally allow for increased yield by further maximizing selectivity and minimizing purge losses (12)

Assuming an oxygen price of approximately $40/ton and a yield of 70% for a commercial EO process, oxygen represents less than 2.5% of the EO feedstock cost. EO is a versatile intermediate used in the production of ethylene glycol and other oxide derivatives such as glycol ethers, polyethylene glycol, polyether polyols, diethylene and triethylene glycols and ethanolamines (13). Most EO producers are integrated with downstream ethylene glycol facilities. World class EO plants with capacities of 0.5B to 1B pounds per year required 400 to 800 tons/day of high purity oxygen. The largest plant in the US is Dow's (formerly Union Carbide) Taft, LA facility with a capacity of 1.5B pounds per year of EO.

Sulfur Recovery Unit (SRU): Hydrodesulfurization of transportation fuels to meet clear air standards produces H_2S that must be treated. The Claus Process, invented by Carl Friedrich Claus in 1883 and further developed by I. G. Farben Industries is the industry standard for recovering elemental S from H_2S. The overall reaction is:

$H_2S + 0.5\ O_2$ ➔ $S + H_2O$

The conversion of H_2S to S takes place in two distinct steps; combustion of H_2S to SO_2 followed by a catalytic reaction of H_2S with SO_2 to form S and water. The primary source of O_2 for the combustion is air; however, enrichment of the combustion air with oxygen has been demonstrated to significantly increase the capacity of an SRU. This is due mainly to higher combustion chamber temperature while avoiding excessive pressure drop that would occur if air flow was increased. For example, enrichment of combustion air to 40% oxygen has been shown to increase SRU capacity by 50% or more (14).

Hydrogen: Large amounts of H_2 are produced and consumed for ammonia and methanol synthesis. In the case of ammonia, the H_2 is typically made by steam methane reforming, followed by mixing with air and secondary reforming to obtain a 3:1 ratio of H_2 with nitrogen. There are also a wide variety of chemical processes that use small to moderate amounts of high purity H_2 (>99.9%).

Refining: Oil refineries include processes that are both H_2 producers (catalytic reforming) and consumers (hydrotreating and hydrocracking). Hydrotreating and hydrocracking are among the oldest catalytic processes used in refining. Originally employed in Germany for conversion of lignite to gasoline, they were later used to process petroleum residues into distillable streams. The first US installation of commercial hydrorefining was at Standard Oil's Baton Rouge refinery in the 1930s. Due to the availability of Middle Eastern crudes following WWII, the use of hydrorefining experienced little growth. Use of hydrorefining processes accelerated in the 1950s; first as a sink for hydrogen available from catalytic reforming, later as means for control the ratios of gasoline and middle distillates (15). Today, hydrotreating and hydrocracking are critical levers available to refiners to deal with the challenges created by the interaction of reduced crude quality, environmental constraints and market demand (15). Hydrotreating removes contaminants (sulfur, nitrogen) and saturates olefins and aromatics to yield a clean product for further processing or sales. Hydrocracking also removes contaminants while converting low value gas oils to higher value products (naphtha, middle distillates, and ultra-clean lube base stocks) (16). As shown in Figure 9, refineries have become net consumers of H_2 over the past 50 years requiring reliable sources of "on purpose" production to achieve high conversion and meet clean fuel specifications (17).

Figure 9. On-Purpose Refinery H_2 Demand

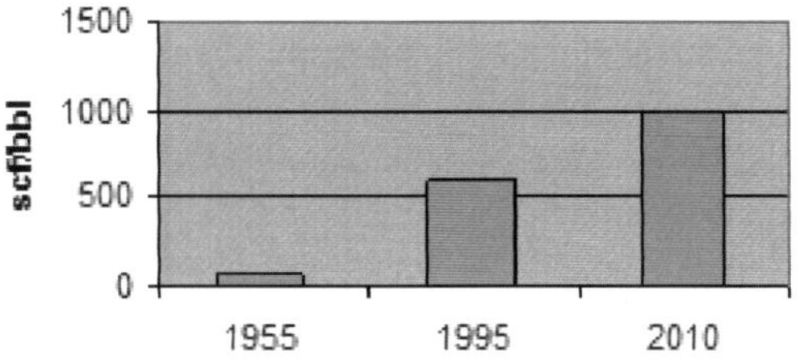

Nitrogen: Nitrogen available as a byproduct from a cryogenic oxygen plant was first employed to create an inert atmosphere by the steel industry in the 1940s. Nitrogen is practically inert at modest temperatures and pressures. Following WWII with the rapid growth of petrochemicals, nitrogen demand increased for blanketing and padding to minimize the possibility of fire and explosions (18). Today, in additional to blanketing, nitrogen is used for mixing, stripping, drying, cooling (liquid nitrogen), emissions control and pressure transfer of liquids. The safety, environmental footprint and efficiency of modern petrochemical and refining operations are all enhanced by the availability of nitrogen.

Nitrogen is also an important fluid used in oil and gas production. Compressed nitrogen is injected down-hole for enhanced oil recovery (pressure maintenance) and gas well fracturing (19).

Emerging Applications

Gasification: Natural gas and crude oil form the foundation of today's petrochemicals industry; however, with prices increases and volatility driven by growing demand and shrinking supplies, commercial chemical and energy projects based on gasification of coal, pet coke and biomass are moving forward throughout the world.

Predating the present natural gas supply systems was the town gas industry, an early commercial gasification technology that developed in Europe and the US in the 1800s initially to supply gas for lighting. Starting around 1900, water gas made by reacting carbon and steam to generate an equal molar mixture of H_2 and CO became available to the chemical industry. Water gas is made by a semi-continuous process where steam is passed over hot coke. Energy must be supplied by periodically flowing air over the coke.

It was not until oxygen was made available in large volumes by the commercialization of cryogenic air separation after 1910 that continuous gasification was possible. Modern gasifiers have their roots with the Winkler fluid bed, Lurgi moving bed and Koppoers-Totzek entrained flow processes developed in the 1920s, 30s and 40s respectively (20). Today, Sasol is the leader in the application of oxygen based coal gasification to produce supply the feedstock for synthetic fuel and chemical processes. Sasol's Secunda complex produces and consumers over 40,000 tons per day of oxygen, mostly for gasification (21).

Gasifiers can be air blown, oxygen blown or enriched air blown. The advantages of oxygen blown gasifiers include lower pre-heat temperature, higher carbon conversion and higher heating value of the effluent gas. For power generation applications, the decision of oxidant supply, air or oxygen, is purely economic (22). For chemical synthesis, the nitrogen free syngas from an oxygen blown gasifier is generally preferred. Single train cryogenic ASUs are under development to supply up to 6,000 tpd of oxygen for gasification applications.

Oxyfuel Combustion for CO_2 Capture and Sequestration: Combustion with oxygen, oxygen lancing and air enriched with oxygen has been employed in a number of industries (glass, steel and cement). The typical benefits include increased throughput, reduced NOx emissions and high thermal efficiency. With many countries moving toward the constraints on CO_2 released into the atmosphere, oxyfuel combustion will play a role as a means to concentrate CO_2 in flue gas thereby facilitating CO_2 capture and sequestration. For example, in the electric power industry, several commercial demonstrations are planned involving combustion of coal with a mixture of oxygen and CO_2 (23). Figure 10 is a schematic of an oxy-coal fired power plant. Oxygen combustion is also being considered for SAGD boilers and FCC regeneration to enable CO_2 capture.

Figure 10.

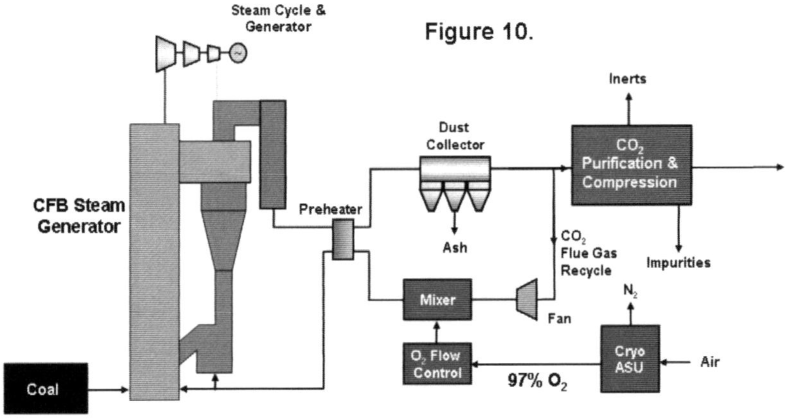

Biofuels: As with petroleum based fuels and products, industrial gases will play a role in efforts to makes fuels and chemicals from renewable biological sources. For example, UOP has introduced a hydrorefining process for vegetable based feedstock. UOP's Ecofining process deoxygenates feed by adding hydrogen to produce green diesel fuel that is superior to biodiesel and petroleum derived diesel (24).

	Petro-Diesel	Biodiesel (FAME)	Green Diesel
NOx Emission	Baseline	+10	-10 to 0
Cetane	40-55	50-65	75-90
Cold flow properties	Baseline	Poor	Excellent
Oxidative stability	Baseline	Poor	Excellent
Energy content	Baseline	Lower	Similar

References

1. Baker, R., *Membrane Technology and Applications*, John Wiley and Sons (2004).
2. Li, J., et al, *Progress towards the Demonstration of OTM Integration with Coal Combustion*, 33rd International Technical Conference on Coal Utilization & Fuel Systems, Clearwater, Florida (June 1-5, 2008).
3. Allam, R. et al, *Cryogenics & Ceramic Membranes: Current & Future Technologies for Oxygen Supply in Gasification Systems*, 4th European Gasification Conference, Noordwijk, The Netherlands (April 11-13, 2000).
4. Isalski, W. *Separation of Gases*; Monographs on Cryogenics 5, Oxford Science Publications (1989).
5. Boyce, Clay et al, *Time for a New Hydrogen Plant?*, Hydrocarbon Processing (February 2004).

6. Morrow, J.M., Corrao, M., and Hylkema, S., *Method of storing and supplying hydrogen to a pipeline,* US Patent US7,078,011 B2 (July 18, 2006).
7. http://www.questairinc.com/pdf/H6200_032008.pdf
8. http://www.uop.com/objects/Polybed_PSA_gas_extr.pdf
9. Gunardson, H., *Industrial Gases in Petroleum Processing*, Marcel Dekker, Inc. (1998).
10. *Industrial Gases in the Chemical Industry*, Linde AG (1999).
11. Lacson, J., *CEH Marketing Research Report – Ethylene Oxide*, Chemical Economics Handbook, SRI International 2003).
12. Papavassiliou; V., Wagner, M. and Day, R., *High purity oxygen for ethylene oxide production*, US Patent 6,040,467 (July 24, 1997).
13. http://www.dow.com/ethyleneoxide/
14. Rameshni, M., *Cost Effective Options to Expand SRU Capacity Using Oxygen*, Sulfur Recovery Symposium - Brimstone Engineering Services, Inc., Calgary, Alberta (May 6th-10th, 2002).
15. Ackelson, D. *Chapter 7.2; UOP Unicracking Process for Hydrocracking; Handbook of Petroleum Refining Processes, 3rd Edition*, Robert Meyers editor, Mc-Graw Hill (2003)
16. Colwell, R., *Oil Refinery Processes: A Brief Overview*, Process Engineering Associates, LLC, http://www.processengr.com/
17. Patel, N., et al, *Creating Value Through Refinery Hydrogen Management*, http://www.h2alliance.com/
18. Kerry, F., *Industrial Gas Handbook – Gas Separation and Purification*, CRC Press (2007).
19. *Get more oil, from more places, more affordably,* Praxair, Inc. (2007).
20. Higman, C. and M. van der Burgt, *Gasification*, Gulf Western Publishing (2003).
21. http://www.sasol.com/sasol_internet/frontend/navigation.jsp?articleTypeID=2&articleId=7400002&navid=4&rootid=4
22. Baukal, C., *Oxygen-Enhanced Combustion*, CRC Press (1998).
23. Shah, Minish, Dante Bonaquist and Richard Victor, *Oxy-Coal for CO2 Capture – Technology Development & Demonstration*, The 33rd Annual International Technical Conference on Coal Utilization & Fuel Systems, Clearwater, Florida, USA (June 1-8, 2008).
24. http://www.uop.com/objects/UOP_ENI_Ecofining_Process%20-%20final.pdf

History of the Development of LNG Technology

Dr. Chen-Hwa Chiu
Senior Technology Advisor
Chevron Energy Technology Company
Houston, Texas

AIChE Annual Conference
Hundred Years of Advancements in Fuels and Petrochemicals
Philadelphia, Pennsylvania, 18 November 2008

Introduction

Liquefied Natural Gas (LNG) started in the USA as a peak-shaving measure. International interest in transporting LNG across the ocean for energy utilization soon developed. Baseload LNG plant capacity has expanded from about 1.0 MTPA to a 4.0 MTPA and recently, up to 7.5 MTPA. The compressor drivers have increased from Frame-5, to single Frame-6 & Frame-7 and now, Frame-9s.

The drivers have evolved from steam to gas and electrical, or a combination of them. There have also been the challenges of using plate-fin heat exchangers versus spiral wound exchangers, coupled with the development of cascade, mixed refrigerant, expander cycles and various hybrids of them.

New energy-saving developments across the LNG value chain include the use of cryogenic liquid expanders, larger LNG storage tanks, and bigger LNG ships with alternate propulsions. New concepts include floating LNG plants, Gravity Based Structures (GBS) and floating LNG receiving terminals. There is also Energy Bridge, reduction of Greenhouse Gas (GHG) emissions, etc.

This paper addresses the development of LNG technology in the transportation and utilization of natural gas in the 21st century [1, 2].

Early Development

Cryogenic industry has its early start since Dr. Carl von Linde developed air and gas separation technologies in the nineteenth century in Munich, Germany. The LNG industry started its early development by using LNG technology for natural gas peak shaving. Peak shaving is a strategy used by the power industry to store natural gas for peak demand that can not be met by their typical pipeline volume. Utility companies liquefy natural gas during low demand and re-gasify it during peak demand to augment available supply.

Cascade cycle was used in LNG plant in the beginning. Later, A. Klimenko presented the mixed refrigerant concept [3] at the LNG-1 Conference. Air Products applied its all mixed refrigerant cycle in the Libya Marsa El Brega LNG plant. Afterwards, Air Products improved it with the propane precooled mixed refrigerant (C3-MR) cycle, which is being used in more than 80% of the LNG plants globally.

Phillips Petroleum invented the cascade liquefaction cycle [4]. This cascade cycle is a closed loop cycle of propane, ethylene, and methane refrigerants. Interestingly, at the time the C3-MR cycle was built in Brunai LNG plant, the cascade cycle was built for Kenai LNG plant in Alaska and Prichard's all MR cycle was built later in Africa. A newer version of Phillips open loop cascade cycle has been built in Trinidad and several other places such as Egypt, Darwin, and Equatorial Guinea.

Early contributors to LNG industry include Lee Gaumer and Chuck Newton who invented the all mixed refrigerant cycle [5] and the C3-MR cycle [6] for Air Products LNG process. The Wilkes Barre cryogenic facility has manufactured the coil wound LNG Main Cryogenic Heat Exchangers (MCHE) since late 1960s. Ludwig Kniel of Lummus invented a cascade cycle and regasification plant synergy with ethylene plant. Ludwig introduced a nitrogen expansion cycle as a subcooling section for LNG process [7].

Dr. C. M. "Cheddy" Sliepcevich pioneered and managed the research, development and implementation of the first commercial process for liquefaction and LNG ocean transport during his work with Chicago Stock Yards and Continental Oil Company, at the University of Oklahoma [8, 9]. For his pioneering research in LNG technology, Cheddy – also referred to as the "Father of LNG", received the Gas Industry Research Award by the American Gas Association Operating Section in 1986 in Seattle. The award, sponsored by Sprague Schlumberger, honored his scientific achievement in LNG research and his contribution to LNG safety. Some of his students, Dr. Hardi Hashemi, Dr. Harry West and Dr. Jerry Havens have further developed his work in LNG safety [10, 11].

Developments in Drivers and Coolers

In the beginning, steam turbines were preferred for LNG plant application because of its prevalence in the oil refineries. Steam turbines were applied at Bontang LNG plant in Badak, Brunei and Das Island LNG plants. Later it was discovered that gas turbines can be more economically applied in LNG plant and therefore new LNG plants started using gas turbines.

Compressor and gas turbine drivers adapted from the electrical industry were the first on the scene. Thereafter, the Frame-5 was redesigned with a split-shaft configuration suitable for mechanical service driver service. This change underlie the shift from GE Frame-5 split-shaft turbine to GE Frame-6, Frame-7 and recently Frame-9 single shaft gas turbines, with power range of 30 to 123 MW. These large turbines are driving large axial and centrifugal compressors. Also, LM-2500 aero-derivative gas turbines have been used in the direct drive application at the Darwin LNG plant in Australia, and the LM-6000 have been used to generate electricity to drive the electric motors which in turn, drive the compressors. This has influenced the application of all electric drivers for more LNG plants in the future.

As the gas turbine drivers are being improved, the water cooled exchangers have been changed to ambient air cooled heat exchangers. This is attributed to two factors: one is the concern over water temperature changes and secondly, due to simple and more efficient usage of large ambient air cooled exchangers.

Heat Exchangers for Liquefaction

Heat exchangers used in LNG are classified into the coil wound heat exchangers and the plate-fin core exchangers.

Coil wound heat exchangers have evolved from smaller sizes to about 15 feet diameter and approximately 200 feet in height and weighs up to 300 metric tons, including thousands of tubing capable of holding internal pressure up to 1,100 psig. Currently Air Products and Linde manufacture this kind of cryogenic heat exchangers, and it can take up to 25 months to complete one exchanger.

Plate-fin exchangers are manufactured by several vendors and are much cheaper than the coil wound heat exchangers. Variations include core-in-kettle exchangers. These exchangers are manufactured by vacuum brazing the aluminum components into the whole exchanger and require shop testing for high pressure performance.

Liquefaction Cycle and Economies of Scale

Phillips Petroleum developed the close loop optimized LNG cascade cycle and improved it in the early 90s to what is known today as the open loop process cycle. For mixed refrigerant cycle - there is the Pritchard Prico cycle, and Air Products all MR and C3-MR cycles. There are also other cycles by some French companies. Conoco Phillips' optimum cascade cycle can be built in large LNG plants up to 8+ MTPA [12]. This process is being used in LNG plants built in Darwin, Egypt and Equatorial Guinea and will be used in the Angola LNG plant.

There have been tremendous developments in liquefaction technology in recent years. In mixed refrigerant cycle, there is the single mixed refrigerant cycle and double mixed refrigerant cycles developed by Shell. Shell also developed the Parallel MR cycle [13], which utilizes the split casing propane compressor arrangement. The Axen's Liquefin cycle, is essentially a dual mixed refrigerant cycle. Air Products has developed the AP-X™ cycle to plant capacity up to 8+ MTPA [14]. Linde and Statoil invented the mixed fluid cascade cycle which is being applied to the Snohvit LNG plant in the Arctic region of Norway [15].

All electric drive configuration is being used in Snohvit LNG to increase overall liquefaction efficiency. Furthermore, cryogenic liquid expanders are now commonly used in liquefaction processes to increase liquid production.

LNG Ships

Newer LNG storage tanks of 160,000 M^3 are in use. In the near future 200,000 m^3 tanks will be deployed above ground. Similarly, larger LNG ships called Q-Flex and Q-Max ships are being built and deployed. The Q-Max has the capacity for about 266,000 m^3 of LNG, with a length of 345 m, a breadth of 53.8 m, and a height of 34.7 m. These new vessels are expected to ship LNG to the US and Europe from Qatar.

Q-Flex LNG ships were previously the world's largest LNG carriers. The Q-Max vessels now feature slow-speed diesel engines that are more fuel and thermal efficient than steam turbine drivers, with a 30% reduction in overall emissions. It also reduces the shipping costs by 30%. Shipping by conventional vessels typically accounts for a third of the LNG price. High efficient propulsion systems and ship board re-liquefaction have helped to achieve further operational efficiencies and a reduction in greenhouse gas (GHG) emissions. 14 Q-Max and 31 Q-Flex LNG ships have been ordered.

Offshore FPSO and Receiving Terminals

There are exciting developments in the offshore liquefaction [16]. Offshore floating LNG studies were performed by Air Products in 1970s. Mobil published their work in 1998 with a moon-pool layout and used the single mixed refrigerant process. Shell is working on Greater Sunrise, offshore Australia and Kudu, offshore Namibia. Statoil and others have worked on Nnwa-Doro, offshore Nigeria, and several Japanese and French outfits have worked on various offshore LNG production schemes. Floating offshore LNG production can eliminate the flaring or re-injection associated with crude production if sufficient gas quantities are available. Recent developments have seen Flex LNG, SBM and Linde, and Teakay working with Mustang on floating LNG concepts.

An offshore LNG receiving terminal was constructed on a Gravity-Based Structure (GBS) and installed offshore Italy. ExxonMobil announced plans to seek regulatory approval for BlueOcean Energy, a floating LNG receiving terminal to supply natural gas to New Jersey and New York.

There are two different locations in America that use the Energy Bridge™ concept. One is the Gulf Gateway and the other is the Northeast Gateway. Excelerate Energy uses fleet of Energy Bridge Regasification Vessels (EBRVs) to dock to these offshore gateways, regasify LNG on board the ship and then transport the re-gasified natural gas onshore by pipelines.

Conclusions

Historically, development of LNG technology has responded to growing LNG demand with new innovation in liquefaction technology, coupled with energy integration of the LNG chain. In the near future, we can envisage an increase in global liquefaction capacity, LNG storage and LNG ship size. The sites for LNG liquefaction plants or receiving terminals will expand to include offshore.

Acknowledgement

I like to thank my Chevron colleagues Bobby Martinez and Ugochukwu Diribe for their useful comments, and to Professor Richard G. Mallinson of the University of Oklahoma for contributions to LNG by Professor Cheddy Sliepcevich.

References

1. S. Huang, C.-H. Chiu and D. Elliott, "LNG: Basics of Liquefied Natural Gas", A book published by *the University of Texas Continuing Education*, Petroleum Extension Service, Austin, Texas, 2007.
2. C.-H. Chiu and C. L. Newton, "Second Law Analysis in Cryogenic Processes", *Energy*, Volume 5, Pergamon Press Limited, 1980.
3. A. P. Klimenko, *First International LNG Conference*, Chicago, April, 1968.
4. J. H. Hughes, US Patent 3,581,510, June 1, 1971. Also J. T. Karbosky and E. A. Harper, *US Patent* 3,408,824, November 5, 1968.
5. L. S. Gaumer, Jr, and C. L. Newton, *US Patent* 3,593,535, December 22, 1969.
6. G. E. Kinard and L. S. Gaumer, "Mixed Refrigerant Cascade Cycles for LNG", *Chemical Engineering Progress*, January 1973.
7. L. Kniel, *US Patent* 3,362,173, January 9, 1968.
8. *The Norman Transcript*, "Endowed OU Professorship Honors Sliepcevich", February 14, 1992. Also, *OU College of Engineering*, "Dr. C.M. Sliepcevich was elected to the Distinguished Graduates Society in 1993", the Board of Regents of the University of Oklahoma.
9. *OKChE, Winter 1988-1989* by School of Chemical Engineering and Material Science, University of Oklahoma: http://cbme.ou.edu/alumni/okchearchive/OKChE_1988_Winter.pdf
10. H. H. West and C.-H. Chiu, "LNG Safety: An Issue of Increasing Importance", *Process Safety Progress*, 24 (3), 2005.
11. J. Havens and T. Spicer, "LNG Vapor Cloud Exclusion Zones for Spills into Impoundments", *Process Safety Progress*, 24 (3), 2005.
12. A. Avidan, W. Varnell and B. Martinez, "Study Evaluates Design Considerations of Larger, More Efficient Liquefaction Plants", *Oil & Gas Journal*, August 18, 2003.
13. B. Pek, E. de Jong, A, van Driel and R. K. Nagelvoort, "Large Capacity LNG Plant Development", *LNG 14 Conference*, Doha, Qatar. March 2004.
14. M. J. Roberts, Y.-N. Liu, J. C. Bronfenbrenner and J. M. Petrowski, "The AP-X™ LNG Process: A Flexible Approach for Reducing LNG Unit Cost", *AIChE Spring Conference*, New Orleans, April, 2004.
15. W. Forg, et al, "A New LNG Baseload Process and Manufacturing of the Main Heat Exchanger", *Reports on Science and Technology* 61, 1999.
16. C.-H. Chiu and L. D. Quillen, "A New Frontier! Offshore Natural Gas Liquefaction", *Gastech 2008*, Bangkok, Thailand, March 2008.

325e Gasification at a Crossroad

Amos Avidan and Lee Schmoe. Bechtel Corporation, 3000 Post Oak Boulevard, Houston, TX 77056-6503

Gasification of coal has been practiced commercially for more than a century. Synthesis gas is one of the fundamental building blocks of the chemical industry, and it is likely to keep its prominence for decades to come. The rate of commercialization of gasification has accelerated in recent years with most of the activity focused in China, and more recently in the US and other countries.

Gasification is used today to convert a wide range of carbon-containing materials into clean synthesis gas. The reactions take place in a reducing atmosphere at elevated pressures and temperatures. Metals and mineral matters in the fuel are melted and recovered as slag. Pollutants, such as sulfur oxides, mercury and particulates are removed from the syngas prior to further processing to chemicals or prior to combustion in gas turbines (IGCC). Gasification today is not able to capitalize on one of its main features - the ability to capture a concentrated stream of CO2, with proven technology and cost-effectively. In a carbon constrained world, this benefit could further increase interest in this technology. High sustained energy prices are further contributing to the use of gasification to convert coal, petroleum coke, biomass, trash, and other carbon-containing materials to power, steam, hydrogen, substitute natural gas (SNG), chemicals, and fuels.

325f Past 100 Years of Refining and Future Expectations

Irv Wiehe, Soluble Solutions, 3 Louise Lane, Gladstone, NJ 07934

The technology of petroleum refining will be compared as it was in 1908 with it is in 2008, a period of time that can be designated as the "petroleum age". In this way the significant developments in refining will be reviewed as well as the technology that has been slow to improve. Using a crystal ball, expectations and challenges for the refinery of the future will be forecast.

100 Years of Petroleum Refining Catalysts

E. Thomas Habib, Jr.
Grace Davison Refining Technologies
7500 Grace Drive, Columbia, MD 21044

100 years ago in 1908, Henry Ford introduced the Model T Ford and ushered in the age when cars were affordable to middle class Americans. At that time petroleum refining consisted primarily of distillation only, and the amount of gasoline required to fuel a large number of automobiles was simply not available. However, a major change was on the horizon for petroleum refining – the widespread use of catalytic processing to greatly improve the yield of fuels. Without these new catalytic processes, the age of the automobile would not have been possible. This presentation briefly reviews the basic catalysts and their historical improvements for the five most important catalytic refining processes: hydroprocessing (hydrotreating and hydrocracking), cracking, isomerization, alkylation, and reforming.

Abstract ID#:
Password:
Session Selection:

327a A Look at the Advances In Catalysis Over the Past Century and Thoughts on the Challenges for the Future

Alexis T. Bell, Chemical Engineering, University of California, Berkeley, 201 Gilman Hall, Berkeley, CA 94720

The discovery and development of catalysts has placed played a major role in advancing technologies for the production of fuels and chemical throughout the 20th century, as well as technologies for the abatement of air pollution from mobile and stationary sources. These advances have, in part, been aided by advances in understanding composition and structure of catalysts and how these properties relate to catalyst activity and selectivity. This knowledge has come as a result of using a wide variety of spectroscopic methods and electron microscopy, together with isotopic tracer methods and measurements of kinetics. Theoretical methods have also advanced to the level where they can provide meaningful insights into the relationships between the local structure of an active site and its activity and selectivity in promoting a targeted reaction. This talk will summarizes the advances that have been made in the field of catalysis and understanding catalytic processes from a molecular perspective and will outline the challenges and opportunities for future research.

327b Turning Surface Science into New Catalysis

Mark Barteau, Chemical Engineering, University of Delaware, 150 Academy Street, Newark, DE 19716

Surface science studies in ultra-high vacuum (UHV) offer the possibility, not only of producing mechanistic understanding of known catalytic reactions, but of discovering new chemistry that may lead to new catalytic processes. The challenge of catalyst development from surface science discoveries is in part one of scale-up. The leap from carrying out reactions on single crystal surfaces in UHV to operation of a typical microreactor is one of approximately 13 orders of magnitude in pressure and at least 4 orders of magnitude in surface area.

We have examined the scale-up of a number of reactions in UHV to catalytic operation. Examples include the synthesis of ketenes from carboxylic acids, reductive coupling of carbonyl compounds, cyclotrimerization of alkynes, and formation of epoxides. In the last of these, we have also utilized results from Density Functional Theory calculations to guide the design of new catalysts. While details vary among these examples, the common lesson is the need to translate the active site, rather than long range structure or composition, from model systems to working catalysts.

327c Selectivity In Surface Chemistry

Andrew J. Gellman, Chemical Engineering, Carnegie Mellon University, 5000 Forbes Ave., Pittsburgh, PA 15213

One of the key contributions of surface science to the discipline of catalysis has been the understanding of selectivity in catalytic surface chemistry. The key feature of surface science methodologies that leads to this contribution is the ability to isolate and study elementary chemical reaction steps occurring on single crystal surfaces. These are two independently important attribute of the surface science method. The ability to isolate and then study the kinetics of elementary steps, decouples their kinetics from those of the overall mechanism of which they are part. The use of single crystal surfaces provides homogeneous environments in which the kinetics are not masked by the influence of surface structure. The impact of these aspects of surface chemistry will be illustrated by three examples: study of transition states to elementary catalytic reaction steps, the study of conformational effects in the desorption of large molecules from surfaces and the observations of enantioselectivity in reactions of chiral molecules with chiral surfaces.

327d Selectivity In Ethane Dehydrogenation Over Chromia

John D. Brooks and **David F. Cox**. Chemical Engineering, Virginia Polytechnic Institute and State University, Blacksburg, VA 24061

The catalytic dehydrogenation of ethane is a potential alternative for the production of ethylene if significant improvements can be made in selectivity compared to the homogeneous free radical chain reaction mechanism inherent with steam cracking. Chromia-based catalysts have found use in the catalytic dehydrogenation of butanes, and have been examined as candidates for ethane dehydrogenation. While C-H bond activation has been established as the rate limiting step, little is known about what controls selectivity over chromia catalysts. The reactivity of different alkyl species (ethyl, ethylidene, ethylidyne) has been examined over single-crystal Cr_2O_3 with thermal desorption.

The observed chemistry will be described in relation to known selectivity issues over supported chromia catalysts.

359a History of Mixing 1908 to 1940

Gary K. Patterson, Chemical Engineering Dept., Missouri University of Science and Technology, Schrenk Hall, Rolla, MO 65401

Innovation to the art and science of mixing from 1908 to 1940 will be reviewed. Discussion will review process design techniques as well as advances in equipment technology.

359b History of Mixing 1940 to 1965

Arthur W. Etchells III, Industrial Mixing, Philadelphia, PA 19106

This period can be considered the beginning of mixing as we know it. It was dominated by a number of individuals who shaped the decades to come.

359c History of Mixing - 1965 to 1985

Richard V. Calabrese, Chemical & Biomolecular Engineering, University of Maryland, Building 090, Room 2113, College Park, MD 20742-2111

The seminal contributions to fluid mixing technology, with respect to both research and application, contributed from 1965 to 1985, will be reviewed and discussed. An overview of how these works have influenced current academic infrastructure and industrial practice will be provided. The contributions that continue to guide future trends will be given the most emphasis.

359d History of Mixing 1985 to 1997

Ronald J. Weetman, RJWeetman, 185 Orchard Dr., Rochester, NY 14618

This talk will discuss the history of mixing from 1985 to 1997. Innovations in impeller and other mixing designs will be discussed. New mixing solutions for needed processes will also be shown. Another great aid that had matured during this period was computational fluid dynamics.

Mixing History
1940 to 1965

Bruce Nauman
Rensselaer Polytechnic Institute

This paper briefly outlines the history of "mixing" from 1940 to 1965. The first decade was dominated by World War II and the subsequent recovery. There were few academic publications, and practitioners remained confused regarding units and correlations for agitator power in stirred tanks until about 1950. The next 15 years, 1950 to 1965, witnessed the birth of chemical engineering science where transport phenomena replaced unit operations and a science of mixing emerged

A 1944 Paper

A Mass Velocity Theory for Liquid Agitation

F. D. Miller and J. H. Rushton
Mixing Equipment Company
Rochester, NY

Rushton's "Kinetic Energy"

$$KE = MV^2/2g$$

$$KE = q\rho V^2/2g$$

$$\therefore KE = A\rho V^3/2g$$

where

M = mass

V = velocity

q = volume per unit time

A = cross-sectional area of stream

g = gravitational constant

The units are screwed up. The equation is really for power, not kinetic energy

$$P_I = A\rho V^3 / 2g$$

but $A \propto D_I^2$ and $V \propto N_I D_I$

Substitute and rearrange to get

$$\frac{P_I}{\rho D_I^5 N_I^3} = \text{constant} = \mathbf{Np}$$

Rushton's "mass velocity theory" gives the power number, but the paper just plots RPM versus HP with different lines for different prop diameters and no data points.

Six years later: Rushton, Costich and Everett, 1950

The D^5 dependence is now recognized and the original publication shows actual data points. The units remain questionable due to apparent confusion between g and g_c.

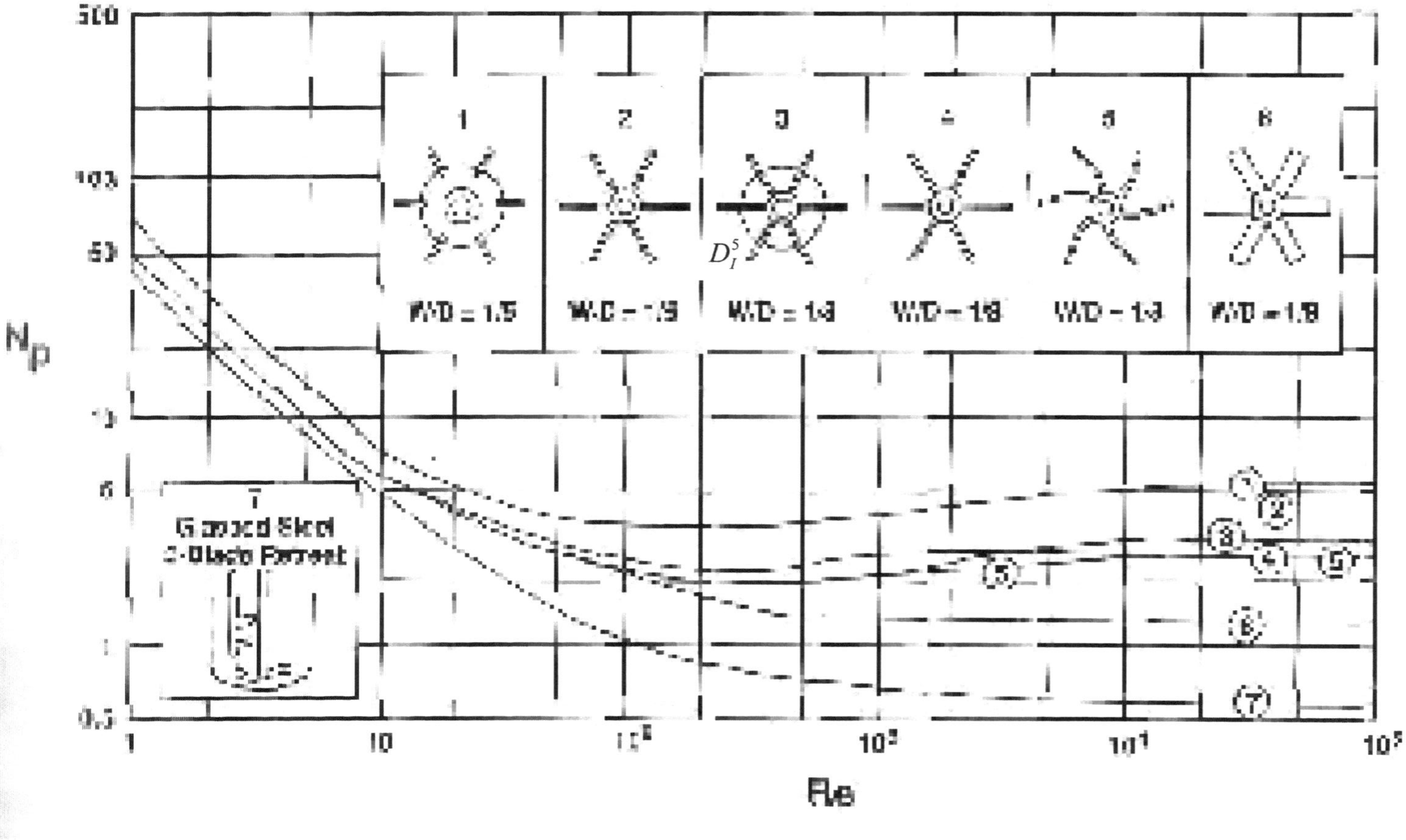

D_I^5

Through and beyond 1965, Rushton and Oldshue dominated the vendor mixing community

- They wrote the annual reviews in I&EC
- Jim began his international outreach

Norwood and Metzner measured mixing times, 1960.

Two key book were being written, but didn't emerge until 1966:

Holland and Chapman

Uhl and Gray

The academic community began an intellectual revolution in 1952 Here are some key players:

Danckwerts, 1952 - 1958

Cleland and Wilhelm, 1956

Mohr, Saxton, and Jepson, 1957

Levenspiel, Smith, and Bischoff, 1957 - 1962

Zwietering, 1959

Bird, Stewart and Lightfoot, 1960

The revolution:

Transport phenomena replaced unit ops

Distributed system models replaced lumped models

1952 Danckwerts starts *Chemical Engineering Science*

1960 BS&L publish *Transport Phenomena*

Danckwert's contributions to mixing

Provided quantitative measures of the extent of mixing

Defined residence time distributions in flow systems

Rediscovered the axial dispersion model and its boundary conditions

Introduced concepts of segregation and micromixing

Cleland and Wilhelm's contributions to mixing

Provided the first numerical solution to a two-dimensional convective diffusion equation, thereby providing the groundwork for CFD

Mohr, Saxton, and Jepson's contributions to mixing

Introduced the concept of striation thickness and showed how it responded to hydrodynamics

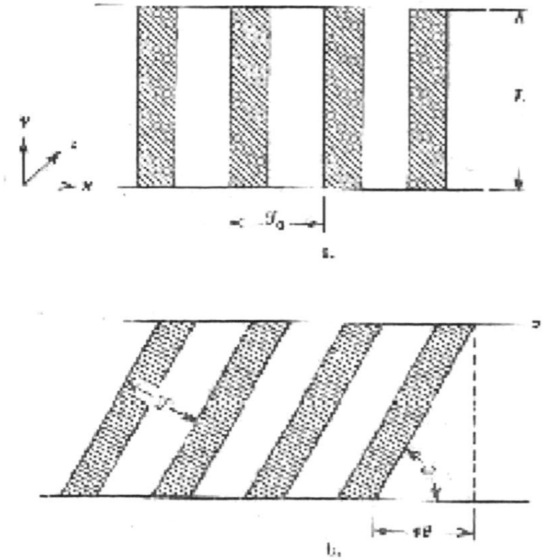

Ranz and Ottino came many years later

Levenspiel, Smith, and Bischoff's contributions to mixing

Developed measurement techniques to characterize pipeline mixing in turbulent flow

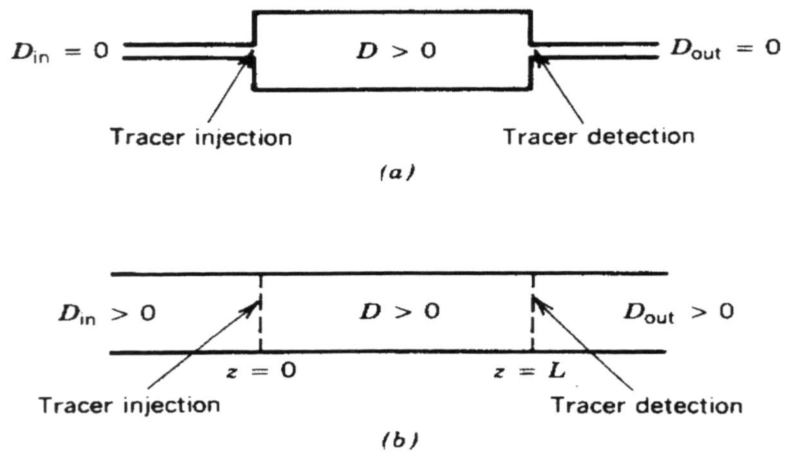

Zwietering's contributions to mixing

Quantified the concepts of segregation and micromixing

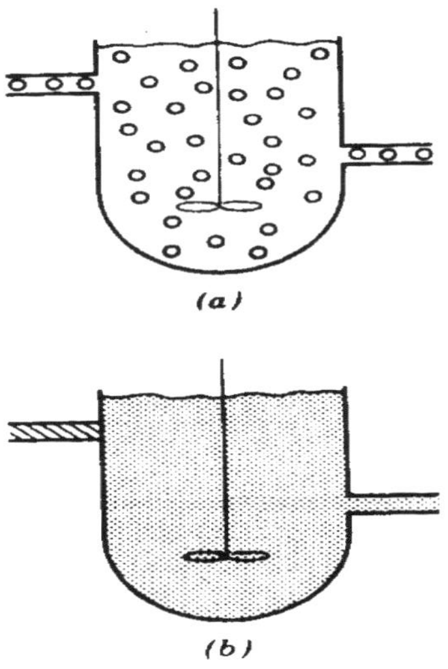

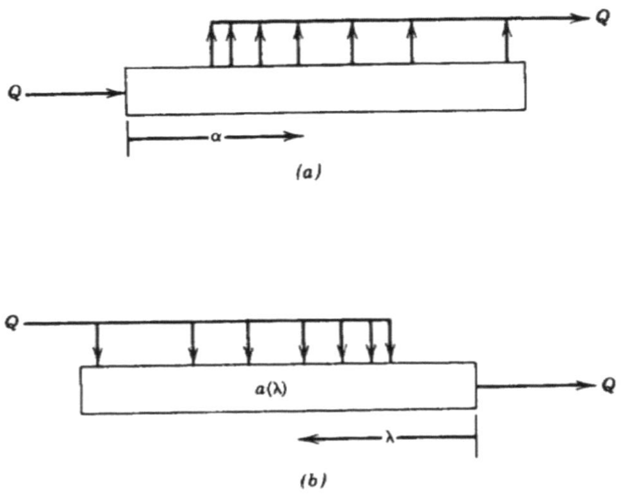

Bird, Stewart and Lightfoot's Contribution to Mixing

Revolutionized graduate education

Professor had to get smarter

Mixing in laminar flow systems could be rigorously defined and even solved

Empiricisms still needed for turbulence

Back to Hardware – Now for Plastics

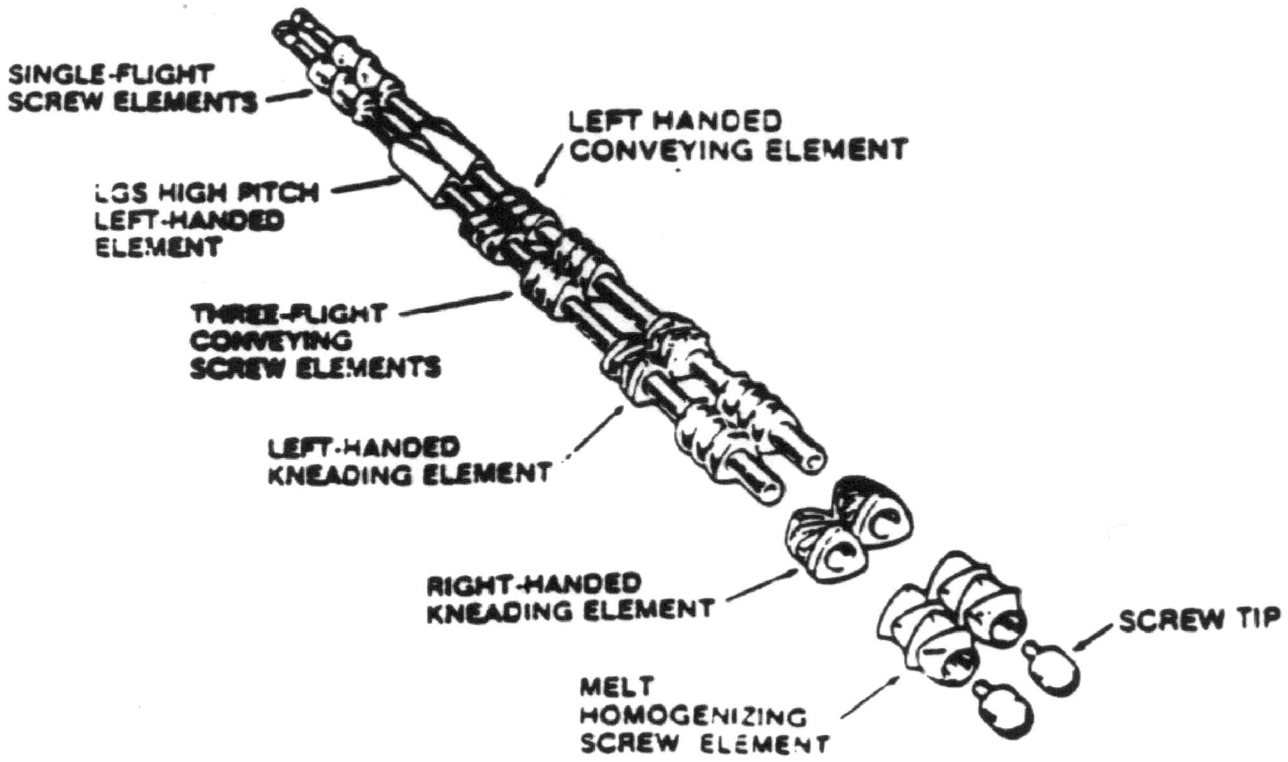

And here is perhaps the first static mixer, circa 1950

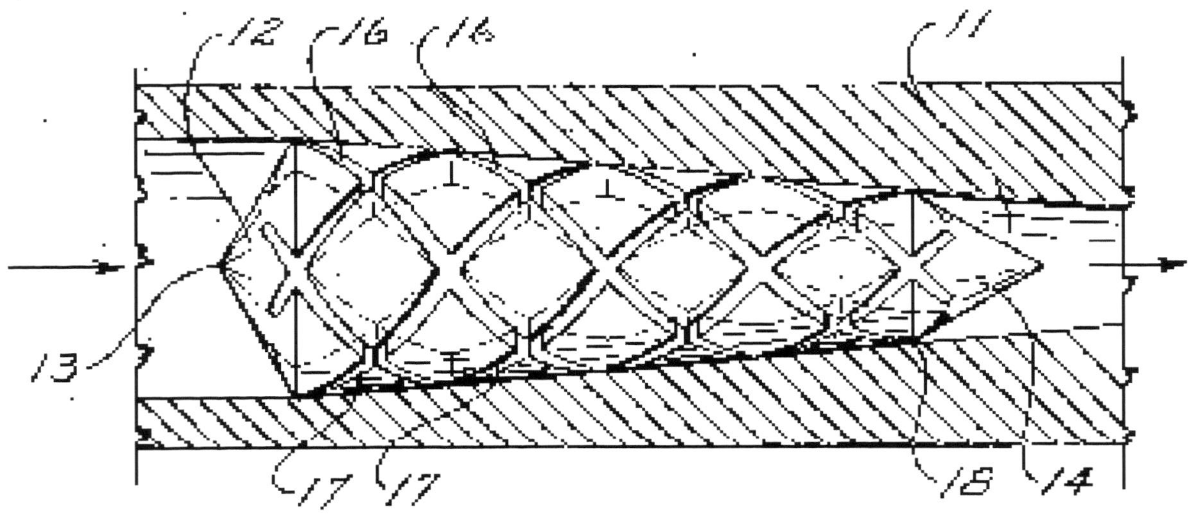

Fig 2

The Greatest Generation?

Except for Kolmogorov, the war years slowed academic research, but the period from about 1950 to 1965 produced remarkable changes in mixing science and engineering.

359f Future Directions of Mixing Research

Richard K. Grenville, DuPont Engineering Technology, DuPont Building D5054, 1007 Market St., Wilmington, DE 19898

Hardware and software developments will change the way mixing equipment is selected and designed for particular process results in the 21st century. Maximizing yields, minimizing waste and optiming energy usage will become important issues that have to be considered in process development and operations. This presentation will dicsuss some of the techniques that will be used to achieve these goals.

WHERE DOES IT GO?
A History of Solid Waste Disposal, Flow Control, and Recycling

David G. Mandelbaum [*]
Brendan K. Collins[**]

1. Waste Management Regulation

Taking waste "out" has been a feature of human settlements as long as there have been human settlements. So, for example, Deuteronomy records a requirement that the Israelites locate latrines outside the encampment and that the army, when in the field, carry with it latrine-digging tools. Deut. 23:13-14. More recently and locally, in 1769 the Pennsylvania General Assembly prohibited distillers, soapmakers, and chandlers from discharging "any . . . foul and nauseous liquor aof any kind whatsoever into or upon any adjacent ground, or into any well, vault, or sink within" Philadelphia or the then-suburbs of Southwark and Northern Liberties. 1769 Pa. Laws chap. 7, § 40. Similarly, section 43 of the same statute provided:

> If any person or persons shall, after the publication her of, presume to cast, carry, draw out, or lay any dead animal or other dead carcase *[sic]* of cattle, sheep, hog, or dog or any excrement or filth from vaults, privies, or necessary-houses and shall leave such carcase, carrion, or filth without buryingthe same a sufficient depth in the ground, on any part of the commons of [Philadelphia and its suburbs], every person or persons so offending and being convicted thereof before any justice of the peace of the city or county of Philadephia [shall pay a fine].

---

* Partner, Environmental Practice Group, Ballard Spahr Andrews & Ingersoll, LLP, Philadelphia, PA. 215.864.8102 / mandelbaum@ballardspahr.com http://www.ballardspahr.com/about/lawyers.asp?id=323.

** Partner, Environmental Practice Group, Ballard Spahr Andrews & Ingersoll, LLP, Philadelphia, PA. 215.864.8102 / collins@ballardspahr.com / http://www.ballardspahr.com/about/lawyers.asp?id=692.

A statute of 1771 made it illegal to throw "rubbish, dirt, or filth" from any carriage or waggon "in any public street, lane, or alley of the city of Philadelphia" unless the location was specifically designated for the purpose. 1771 Pa. Laws chap. 7, § 30.

In the early 19th century, the Philadelphia Waterworks were installed at the base of Fairmount, the hill at the top of which the Philadelphia Museum of Art now sits. Initially, steam engines pumped water from the Schuylkill River to the top of the hill, and it flowed by gravity into the city to the east. Later, the dam was installed, and water power pumped clean water to the top of the hill. An 1803 ordinance prohibited disposing of "any kind of filth" into the basin and canal feeding the pumps. Ultimately, the City established Fairmount Park -- the largest urban park in the United States -- to protect the purity of the water supply.

An ordinance of 1818 authorized the Philadelphia Board of Health to enter into a house to correct an offensive privy. An ordinance of 1878 specifically addressed solid waste transportation providing that:

> no person or persons shall collect or remove kitchen garbage and offal from any dwelling, hotel, restaurant, or any of the streets, alleys, lanes, courts, or alleys of the same except though be collected or removed in water-tight carts, wagons, or other vehicles, and securely covered so that none of the contents shall fall, leak, or spill therefrom or be exposed to public view . . . .

Phila. Ordinance of Dec. 31, 1878, § 1.

These statutes and ordinances presage more modern solid waste management regulation. In essence, they all elaborate on the common law of public nuisance. They prohibit actions or maintenance of conditions that (a) interfere with a water supply or (b) cause offense by odor or otherwise. What has changed over time is our ability to detect interferences with water supplies and our sensitivity to offenses.

The Refuse Act of 1899, *as amended*, 33 U.S.C. § 407, for example, prohibited deposition of solid material -- as opposed to any liquid sewer discharges -- to navigable waters of the United States. State statutes began to impose controls on upland disposal of wastes in the mid-20th century. For example, in Pennsylvania, Act 241 of 1968 first enacted the Solid Waste Management Act, which was later repealed and replaced by Act 97 of 1980, Pa. Stat. Ann. tit. 35, §§ 6018.101 to .1003. At the federal level, Congress first enacted the Solid Waste Disposal Act in 1976, *as amended*, 42 U.S.C. §§ 6901-92k. Those federal amendments primarily focused on hazardous waste management through the Resource Conservation and Recovery Act in 1979 and the Hazardous and Solid Waste Amendments of 1984.

Mid-century solid waste management regulation primarily addressed disease and odor vectors. The federal statute prohibited the maintenance of an "open dump." 42 U.S.C. § 6945. The Environmental Protection Agency promulgated criteria for classification of facilities as "open dumps" and practices as "open dumping." 40 C.F.R. pt. 257, subpt. A. While the federal government never brought much enforcement litigation under those provisions, some private citizen suits were successful at enforcing them. *See, e.g., O'Leary v. Moyer's Landfill, Inc.*, 523 F. Supp. 642 (E.D. Pa. 1981). The open dumping criteria emphasized regular cover, litter control, the avoidance of open fires, and discharges of leachate to surface waters.

Discharges of leachate to *surface* water would typically require a permit under section 402 of the federal Clean Water Act, 33 U.S.C. § 1342. So, for example, when the State of New Jersey, various municipalities, and certain environmental groups sought to force modification to the City of New York's marine solid waste operation by which the City transported solid waste from the four other boroughs to Fresh Kills Landfill Staten Island by barge, losing tons of solid waste in the process, they invoked the Clean Water Act in *Township of*

Woodbridge v. City of New York, No. 2:79-cv-01060-FSH (D.N.J.). That theory still survives. *Reynolds v. Rick's Mushroom Service, Inc.*, 246 F. Supp.2d 449 (E.D. Pa. 2003).

At the state level, state statutes have prohibited maintenance of conditions on land that cause pollution or a threat of pollution for a very long time. In Pennsylvania, those portions of the Cleans Streams Law, *see esp.* Pa. Stat. Ann. tit. 35, §§ 691.316, .401, were enacted in the 1930s to address discharges from coal mining. *See, e.g., Commonwealth v. Barnes & Tucker Co.*, 319 A.2d 871 (Pa. 1974).

Federal law did not so clearly address groundwater contamination. Landfill standards sought to protect groundwater resources using engineering controls. However, until the late 1970s, soils seemed to be very good at filtering landfill leachate and protecting groundwater quality. Certainly soil stops a can from migrating downward. It also seemed to filter metals -- the contaminants of concern at the time -- because metal ions would bind to the soils. Volatile organic compounds, which are non-polar and therefore do not readily bind to the soils, could not be detected at concentrations below parts per thousand. Accordingly, they did not appear to be present. Landfill design standards relied upon "natural renovation" technology, requiring a certain depth of soil between the bottom of the waste and the top of the water table. Only in the late 1970s did regulations change to require lining of non-hazardous waste landfills. *See Franklin Township Bd. of Supervisors v. Dept. of Envtl. Resources*, 1992 Pa. E.H.B. 266.

More recent regulations have mandated both lining and capping non-hazardous waste landfills under part 257 of title 40 of the Code of Federal Regulations. That presents a problem for landfill designs that rely upon degradation to reduce volume or that seek to exploit production of methane gas.

In addition, more recent additions to the solid waste regulatory scheme have addressed waste that is offensive for reasons other than odor, vectors, or water contamination. For example, in the 1980s, but the federal and state statutory schemes began to regulate medical waste. While this waste in fact has lower microbial content than much municipal solid waste, it frightened many in the wake of beach washups during the summer of 1988. Subchapter X of the Solid Waste Disposal Act established a demonstration medical waste tracking program. 42 U.S.C. §§ 6992-92k. States adopted statutes like the Pennsylvania Infectious and Chemotherapeutic Waste Disposal Act; Pa. Stat. Ann. tit. 35, §§ 6019.1-.6.

As waste management regulation has become more sophisticated, regulators have identified wastes in an intermediate status between clean fill and municipal solid waste, like construction and demolition debris. Other wastes have been regulated as intermediate between municipal solid waste and hazardous waste. In Pennsylvania, all non-hazardous industrial waste is specially regulated as "residual waste." Wastes containing asbestos or polychlorinated biphenyls are also given separate and special treatment.

As the result, the non-hazardous, solid waste management regulations applicable to generators, transporters, and disposal facilities fill hundreds of pages in Pennsylvania. Detailed rules can give rise to significant enforcement. Rules can conflict with sensible operations. To take one example, in Pennsylvania, putrescible waste may not be stored for more than 24 hours except over a weekend in a transfer facility. That typically requires the facility to be clean to the floor before closing for the night. But every large municipality requires at least one transfer facility that remains open 24/7, or else street cleanups, hospitals and other 24-hour institutions, and similar operations cannot obtain proper service. Assuring that a facility that never closes is clean to the floor once daily can prove to be a difficult task.

While in any given case the original rationale for any of these rules may seem lost, in almost every instance, the regulations seek to do the same things that the 18th century rules sought: to protect water supplies and to guard against offensive conditions. The difficult task ahead is to revise the now complicated rules to accommodate disposal technologies that rely upon degradation, and do not seek solely to isolate waste from the environment in a plastic liner.

2. Waste Flow Control

The earliest forms of municipal regulation of waste disposal were, in fact, nothing more than what we would recognize as waste flow controls -- laws designating the locations where citizens were required to dispose of their refuse. Three hundred and fifty years ago, in New York City (then, of course, New Amsterdam), "[t]he burgomasters and schepens ordained that all such refuse be brought to dumping-grounds near the City Hall and the gallows or to other designated places." Maud W. Goodwin, *Dutch and English on the Hudson* 105 (1919). While it is ordinarily favored in modern times to haul garbage *away from* City Hall rather than *towards* it, there are numerous examples that demonstrate that waste disposal regulation was at its very inception not about *how* waste was disposed, as it largely is today, but rather about *where* waste was disposed. The inevitable nuisance posed by a growing population's waste disposal was managed by assuring that it was concentrated in certain locations, selected either for their isolation or, as in the case of Philadelphia's artificial riverfront, to achieve some civic purpose.

In the modern era, the focus of waste regulation is primarily on the manner of disposal, with less emphasis on the location of the disposal. This transition was logical; since the only purpose of regulating disposal was to ensure that the nuisance posed by waste was minimized. Environmental protection regulations came into effect not only prescribing technical standards, such as landfill liners, but also mandating certain disposal practices, such as recycling.

These environmental regulations, however necessary and well-intentioned they may have been, had two important effects on the market for solid waste disposal services. First, they created very significant price differentials between disposal sites. New facilities, be they landfills or "resource recovery facilities," were (and remain) very expensive to build, and their owners were compelled to charge much higher prices to recover these higher costs. Yet, new solid waste facilities were required to address a growing capacity crisis in many regions. In many cases, purchasers of disposal services could choose between older, grandfathered landfills at a very low cost, and newer environmentally compliant facilities at a much higher cost. Private purchasers overwhelmingly chose the cheaper option, leaving the new facilities without an adequate customer base to support operations and more importantly, to make loan payments.

Second, a combination of market stimulation from government incentives and reawakened Yankee frugality fostered the recognition that solid waste, for all of its negative characteristics, actually possesses value in the form of recyclable materials, especially aluminum and other metals. While this value may not be enough to offset the cost of collecting the waste, removing the recyclables and disposing the residue (in which case we should all be paid for allowing our waste to be collected), a waste hauler proficient at extracting this value, and at the same time reducing the volume of waste to be landfilled, could claim for itself a meaningful economic benefit.

In the late 20th Century, then, waste flow control reemerged not as a technique for confining waste disposal nuisances in a "trash ghetto," but as a means to assure that new solid waste facilities, as expensive as they were to build and to use, would have an adequate supply of waste to generate the revenue necessary to pay for the facilities, and to sustain operations. In a typical arrangement, a municipal sponsor of a new waste facility would issue bonds to finance the construction of a facility, and lend the bond proceeds to the project developer, who would build the facility. The revenue generated by the facility's operations would be pledged to repay the bonds. In order to guarantee that the revenue stream would be sufficient, the municipality would commit to adopt a waste flow control ordinance requiring disposal of all waste generated within the municipality to the new facility.

Not surprisingly, waste haulers, competing disposal facilities and other groups often objected. In 1993, the case of *C&A Carbone, Inc. v. Town of Clarkstown (NY),* reached the Supreme Court of the United States. Opponents of waste flow control argued that the market for waste disposal transcends state boundaries, and therefore, under a legal doctrine known as the "dormant Commerce Clause," can only be regulated by the federal government. Therefore, they argued, municipalities, and even states, lack the constitutional authority to adopt waste flow control regulations. In its landmark decision, the Supreme Court agreed, ending waste flow control as a tool for locking up waste supplies, jeopardizing the assumptions of waste facility financiers across the country, and in some cases, plunging brand new facilities into bankruptcy.

Solid Waste Landfills
The Road to Green Energy

Introduction

Disposal of solid waste, liquid waste and the ultimate in waste management – burial of the deceased, has been a societal problem since early man first dropped his knuckles while moving upright on two legs. Man learned to separate disposal from community living very early, but it was millennia before the impact of waste disposal on diseases and its transmission became common knowledge. While there was some suspicion of a relationship, it was not until the development of the microscope in the 17th century that the first cause/effect relationships were observed. Scientific progress was slow, and it required another 200 years, until in the middle of the 19th century, an action as simple as removal of the handle from a shallow and contaminated well on Broad Street in London effectively reducing the spread of cholera, dysentery and other water-borne diseases.

It was clear that the relationship between water-borne contamination and disease transmission was established, but not so clear was any relationship with solid waste disposal. Throughout the second half of the 19th century and the first half of the 20th century, public health emphasis was on water-borne mitigation of disease. Solid waste was considered more of a nuisance problem, but the combined impact of liquid and solid waste was impetuous for the passage of the 1899 Rivers and Harbors Act. After 1900 the rate of change for disposal of solid waste accelerated and Table No. 1 is a summary of the pertinent laws and regulations promulgated by the Federal Government or the Commonwealth of Pennsylvania. Regulatory involvement in other states followed a similar calendar, and Table No. 1 will be used as the basis to discuss their "Road to Green Energy".

Table No. 1 – Milestones in Solid Waste Regulations

Year	Regulation
1899	River and Harbor Act
1963	Clean Air Act
1965	Solid Waste Disposal Act
1968	State of PA Act 241
1969	National Environmental Policy Act
1970	Resource Recovery Act
1970	Occupational Safety and Health Act

1972	Clean Water Act
1976	Resource Conservation and Recovery Act
1978	Public Utility Regulatory Act
1980	Comprehensive Environmental Response, Compensation and Liability Act
1980	State of PA Act 97
1987	State of PA Act 101
1995	Section 29 Tax Credits
2002	State of PA Act 90

What Energy?

In the 18th and 19th century it was known that anaerobic digestion of waste, both human and animal, would produce off gas that could be used as fuel. Methane digesters were widely used on farms for lighting and as fuel for pumps and other equipment before the massive push for rural electrification in 1932. When electricity became available, use of methane fuel ceased to play a significant role. Prior to World War II solid waste disposal was a mixture of three methods: Incineration in the larger metropolitan areas or ocean dumping for those near the coast. Open burning dumps, many in swamps and wetlands Industrial on-site burning or burial. Few restrictions existed for solid waste disposal, except for control of nuisances, until the second half of the 20th century. The concept of solid waste disposal in a sanitary landfill, i.e. no burning and the need to cover waste to control fires, flies and other vectors only evolved about 1950, although the concept was known for decades prior to that time. The amount of gas generated by a landfill was not known to be a problem because both open burning and absence of cover allowed gas to escape to the atmosphere and burning the reduced quantity of waste suitable for methane generation. After the number of operating sanitary landfills increased so significant numbers after 1960, gas generation was identified as a major factor in the design and operation of safe landfills.

1899 Rivers and Harbors Act

The 1899 Rivers and Harbors Act was passed to specifically address the problem of jetsam and flotsam in the waters that were used for transportation and shipping. Without reasonably available solid waste disposal, removal of accumulated trash and street waste (think horses) was simply washed into nearby rivers and bays. At many locations this waste also included domestic sewage or at best wastes that received primary treatment only. One must remember that this was still more than a quarter century before Streeter and Phelps postulated their seminal equation that linked the quality of water with waste loading. As with most early laws in addition to floating waste, odor was the principal nuisance over which control was sought.

An interesting result of the Act was the requirement for harbor and port operators to site disposal landfills for items removed from the water. Many of these landfills ultimately served both for disposal of waste removed from the water as well as for the waste generated from the surrounding city or municipality. Marsh land and flood plains in proximity to the water front were widely used for waste disposal. Most of these landfills were open dumps and no effort was made to cover or control public health nuisances. As result no gas was generated nor any energy recovered from the open burning of waste deposited for disposal.

1900 to 1960

For the next sixty years numerous environmental regulations were passed to control smoke (air pollution) and both industrial and domestic wastewater. In Pennsylvania water and waste water acts were passed in 1907, 1912, 1937, 1945, but none of these had any significant impact on solid waste disposal. The first act that related directly to solid waste in streams was passed in 1955, and it related to coal fines discharged into the waters of the Commonwealth. Sedimentation dams and dredge spoil fills were common, but these were designated monofills and joint disposal with household and commercial waste was not practiced as was the case with 1899 Rivers and Harbors Act.

Clean Air Act of 1963

With the passage of this Act, the first regulations passed by the Federal government were implemented. This legislation was without real direction, so that local and state governments passed their own implementing legislation. Strong regulations were passed in California as well as in major U.S. cities including Philadelphia, New York St. Louis plus almost every major California city. Stunned by the quick and comprehensive local government response, the U.S. passed a much stronger and comprehensive Law in 1970. This Act was reauthorized in 1977 and 1990 and the most significant result of this law was the development of the New Source Performance Standards (NSPS) for specific industries. Subsequently standards were adopted for Waste-to-Energy plants as well as gas emissions from landfills. These regulations are known as the MACT or Maximum Achievable Control Technology. These MACT regulations have imposed significant influence on how solid waste is managed, and more importantly upon the cost for treatment and disposal.

MACT regulations for sanitary landfills require a landfill to control the emissions of all gases, dominantly methane and carbon dioxide, when the weight of emission exceeds 1000 metric tonnes. A collection system is required to channel all gas into a system for treatment or use of the gas in a manner that facilitates energy recovery. Simple disposal, i.e. thermal destruction in a flare requires a destruction efficiency of 95 percent. As the value of energy increases, numerous ways have been employed to recover that energy. Table No. 2 is a list of energy recovery systems in common use.

Table No. 2 – Energy Recovery Systems

Typical Systems used for Landfill Gas Energy Recovery
Direct use in Boilers, Space Heat or Industrial Furnaces
Electrical Generation in IC Engine
Prepare Landfill Quality Gas (Remove non-combustibles)
Prepare for Use in Vehicles
Use as Feedstock to make Methanol

The 1990 Law added control of toxic air pollutants to the scope of air management goals. For landfills, trace organics that were listed toxic constituents, had to be measured and eliminated consistent with *de-minimus* emissions. The threshold was set at 10 tons per year for a single constituent and 25 tons per year cumulative for all constituents. Even for concentrations in the PPMV range of a toxic constituent, a large landfill could be snared into air toxic compliance requirements.

Law of Unintended Consequences
Throughout the evolution of laws and regulations to "fix" one problem resulted in the creation of an unintended one. The first of these was the Environmental Policy Act of 1969, the Clean Air Act of 1970 and the Clean Water Act of 1972. To a lesser extent the 1899 Rivers and Harbors Act also had some influence on waste disposal.

Environmental Policy Act of 1969

This Act was the first set of requirements that moved beyond nuisance mitigation to humans and encompassed both land and water habitats. Alteration to the natural environment required total front-end planning and a projection of what impacts would result on the terrestrial and aquatic environments. The set of requirements eliminated disposal in marshes and wetlands well before the impact of leachate on groundwater quality was scientifically established. As recent as 1970 Dr. Roscoe Kandle, New Jersey Commissioner of Health, extolled the virtue of filling low areas with waste and the elimination of marshes that are breeding areas for mosquitoes. With one regulation the siting of a landfill was completely altered from the use of marginal land to use of higher value land.

Clean Air Amendments of 1970

Solid waste professionals frequently point to the potpourri of clean air requirements as an example of the implementation of requirements without the considering the consequences of the impact on either the land or the water environments. The early regulations emphasized removal of particulate matter, and disposal of the removed PM often posed problems. Dry systems generally relied upon land disposal, mostly in monofills, but which were often dusty and dust nuisances. Wet systems, especially small wet scrubbers often discharged into municipal sewer systems greatly increasing the solid loading on the POTWs as well as substantially increasing maintenance on gravity systems and pumps.

Solid Waste Disposal Act of 1965 (PL 89-272)

The first Federal act that directly identified solid waste for regulation and control was very broad and did not contain any enforcement sections. It was the position of the federal government that solid waste regulation was the responsibility of the states and each should develop its own regulatory and enforcement laws and regulations. This Act did recognize the need for research and development and pursuant to the implementation more than 60 projects were funded. Included in the list of funded projects were the first in-depth studies of leachate and gas generation from sanitary landfills. Until this point in time information on sanitary landfilling was derived from landfills and investigations in the State of California. Because of the arid climate, both gas and leachate were not considered to be significant problems. As a result of the expanding use of the sanitary landfill method into wetter climates, the magnitude of potential leachate and gas generation problems became clear. This Act was the forerunner to the 1970 Resource Recovery Act and the 1976 Resource Conservation and Recovery Act, better known as RCRA.

Pennsylvania Act 241 (1968)

The Pennsylvania Solid Waste Management Act was the most significant single act that moved the state toward energy from landfill gas. Regulations that implemented this Act proceeded to enforce closure of open dumps and malfunctioning landfills. The State proceeded to close more that 1800 landfills and open dumps. Secondly the Sate mandated that all municipalities having a population density greater than 300 persons per square mile must prepare a 10 horizon solid waste management plan. The third result of Act 241 was the first set of rules and regulations that guided both construction and operation of sanitary landfills. Grants and general prodding focused planning at the county level of government, and as a result most planning districts were for a single county. The result of this planning activity was the permitting and construction of larger regional landfills that were of sufficient size to generate landfill gas at a rate where recovery today is both feasible and fiscally attractive.

Research and Development conducted with funding under the 1965 Solid Waste Disposal Act evolved the model (now the HELP Model) that predicted the adsorption capacity of municipal refuse, and let to an experimental period in the 1970 decade when liquid waste could be placed in a landfill under the assumption that no leachate would be generated provided that the rate of vertical filling exceeded liquid plus precipitation inflow. California landfills were already making liquid placement in landfills a common practice, and for a short period of time Pennsylvania allowed this practice in landfills that has impermeable liners to collect and treat leachate.

1972 Clean Water Act (PL-92-500)

The Clean Water Act, while focusing upon discharges of liquids to surface and ground water, did exert some significant influence upon the practice of sanitary landfilling. The first impact was the expansion of the number of facilities that were constructed to provide service to small and large municipalities. The result of the increased number of treatment plants was generation of both raw and digested sludge, mulch of which is deposited in landfills. Sewage sludge, especially raw sludge provides moisture and nutrients that increase and accelerate the generation of methane gas in landfills.

A second and not so benign impact was the diversion of liquid waste into landfills when municipalities developed ordinances to regulate the pretreatment of industrial wastes. With no disposal restrictions on the liquid and semi-liquid wastes, landfills received a wide array of industrial wastes including solvents, pesticides, herbicides, hospital waste, and other potentially deleterious wastes. Unlined sites receiving these types of wastes often were added to the CERCLA or Superfund list in the 1980's.

1976 Resource Conservation and Recovery Act (RCRA)

RCRA or PL 94-580 was the most comprehensive law written to address the management of wastes upon the land. RCRA addressed with both hazardous and non-hazardous waste treatment, storage and disposal. Most widely used methods of management waste upon the land, including landfills, lagoons, land farming, and deep well injection became regulated. The law was subdivided into hazardous (Subtitle C) and non-hazardous (Subtitle D) requirements.

The Subtitle C regulations were extremely broad and specified control over not only disposal, but also storage of drums, storage in tanks, treatment of wastes including incineration and all recycling, recovery and reuse operations All waste that were transported required transporters who were permitted and a comprehensive manifest system was used to monitor the transport and disposal of all hazardous wastes.

For non-hazardous wastes that were disposed of on the land, EPA was required to evaluate landfilling, and to prepare a criteria that could be used to classify a landfill as either acceptable for sanitary landfilling or to classify the facility as an open dump. After development of the criteria for classification as an open dump, a national survey identified almost 40,000 facilities that were substandard and were classified as a dump.

Hazardous and Solid Waste Amendments of 1984 (HSWA)

The 1976 RCRA law was vague relative to land disposal of waste and temporary landfill regulations were promulgated in 1982, pending passage of additional legislation. In 1984 the amendments to RCRA were passed and regulations that effectively changed design of both Subtitle C and D landfills became law. Requirements for use of composite liners, leachate monitoring and gas control were part of the new legislation. HSWA also banned disposal of liquids in landfills, and phased out these practices for period of several years. Regulations also identified constituents of concern in leachate and provided a list for testing. Subtitle D regulations essentially rendered the trench method of landfill operation unacceptable. In 1991 RCRA was again amended, and the principle focus of the amendment was to define the conditions of "financial assurance". Post closure care was defined as no less than 30 years, and sufficient funds must be provided to operate the maintenance of the landfill during that period of time.

National Energy Act of 1978

As a result of the energy turmoil that persisted throughout the 1970 decade, Congress passed the National Energy Act in 1978. This Act consisted of five bills addressing different segments of energy policy.

1. Natural Gas Policy Act of 1978
2. Power Plant and Industrial Use Act of 1978
3. Public Utilities Regulatory Policy Act of 1978
4. National Energy Conservation Policy Act of 1978
5. Energy Tax Act of 1978

Of these acts the most relevant to landfill gas is the Public Utilities Regulatory Policy Act (PURPA), PURPA opened the electrical generation and distribution system to deregulation of utilities and opened service areas to competition, and also required utilities to purchase electricity from non-utility generators. The rate that the utility was obligated to pay was based upon a concept of "avoided costs", a rate which was usually well below the market price.

The second portion of the National Energy Act is the Energy Tax Act that provided tax credits, investment credit and depreciation advantages for renewable sources of energy.

Comprehensive Environmental Response, Compensation and Liability Act (CERCLA Or Superfund)

Superfund regulations imposed remediation criteria on many sanitary landfills, especially those that caused ground water contamination from landfill operation. One condition in CERCLA exempted non-containerized gas (landfill gas) from remediation and focused requirements for gas management upon air quality regulations. However, for landfills that were included on the list for which remediation will be required, development of landfill gas recovery and utilization programs was subject to another level of oversight and approval.

Pennsylvania Act 97 (1980)

In Pennsylvania, like many of the more populated states on the east coast, waste disposal was becoming a major issue in the late 70's. The Solid Waste Management Act required the Department of Environmental Resources (DER) to develop a state wide solid waste plan and further develop regulations to manage waste. The DER then asked local governments to develop solid waste plans. For the first time landfill developers were required to post Bonds or other forms of surety accepted by the state to cover closure, post-closure and remedial costs. The planning and regulations lead to increased reliance on regional landfills and in California landfill gas for beneficial use had begun.

Pennsylvania Act 101 (1987)

This Act is often thought of as Pennsylvania's start of mandated recycling. It did require all of its municipalities with a population of over 10,000 to recycle at least three items from a list of eight. This Act for the first time banned non-hazardous items from municipal landfills – tires, lead acid batteries and loads primarily composed of leaf waste. There was an extensive recycling grant program started and funded by a $2.00 fee on every ton of waste disposed of in a Pennsylvania landfill. There were new oversight and costs from Act 101 for every landfill. There was a minimum $1.00 per ton "Host Fee" required for "Host Municipalities". Landfills were now required to offer testing of drinking wells for all contiguous property owners. Host Municipalities could now have "Host Inspectors" visit the landfills as often as they wanted and the Host Municipality would receive 50% reimbursement for these costs from the state. As a result of more regulations and costs the trend toward larger regional landfills accelerated as Act 101 was implemented. Many of large national waste companies that owned landfills were beginning to see landfill gas beneficial use projects as a steady source of revenue, tax credits and positive public perception that would go on long after a landfill was closed.

Section 29 Tax Credits (1995)

If Agreements were executed with a third party by the end of 1995 there were tax credits available for beneficial use projects. These tax credits spawned one of the biggest development periods in the history of landfill gas green power.

Pennsylvania Act 90 (2002)

This Act is unique to Pennsylvania and was a legislative response to being the biggest disposer of out of state waste in the country. Pennsylvania already had the most expensive fee structure at its landfills in the country and this Act added $4.00 per ton. The extra cost of disposal in Pennsylvania lead to the development of large regional landfills in neighboring states. Most of the out of state waste disposed of in Pennsylvania is from New York City and New Jersey. In some of the neighboring states rail transfer stations have been developed to decrease the transportation costs through and around Pennsylvania. Many of these new "Green Field" landfills had landfill gas to energy projects from the beginning to help pay for and help sell the landfill to regulators and the local public.

Summary

As the cost of energy doubles or triples prices in very compressed periods of time, and the use of "green" energy becomes a high priority in the national psyche, landfill gas becomes a readily available renewable fuel that can fill a small niche. Landfill gas utilization has many positive attributes such as:

- It is renewable
- Development costs are competitive with fossil fuel acquisition
- Utilization can occur at the source
- Capture and destruction is already required
- Capture and use can be independent of the energy grid

Some negative aspects include:

- Well field management can be complex
- Production may have short term life
- Some pretreatment may be needed for selected systems
- Multiple regulatory agencies (air, solid waste, waste water, OSHA) will have oversight.

It must be remembered that a landfill is first and only a facility for disposal of solid wastes. Landfill gas is a result of the disposal function, and must be managed according to the prevailing regulations relevant to landfill permits. Any gas program used for energy recovery must fit within that landfill operation framework.

368c The Evolution of Hazardous Waste Law and Regulation

Mary Ellen Ternes, McAfee & Taft, 10th Floor, Two Leadership Square, Oklahoma City, OK 73102-7103

This presentation will take the attendees from the history of solid waste regulation to regulation of wastes that are more hazardous. This discussion will review initial attempts to protect the public health and environment from hazardous substances through the period after World War II, when the volume and toxicity of waste streams from both industry and municipalities spiked with industrial growth and the robust development of synthetic chemical manufacturing, and then to the evidence of systematic mismanagement which gained national attention in the Niagara Falls, New York residential community known as "Love Canal" where residences floated on industrial sludge, followed by Times Beach, Missouri where horses died from exposure to used oil contaminated with dioxin sprayed on roads to control dust and then, of course, the infamous "Vally of the Drums" outside of Louisville, Kentucky.

The discussion will continue with the development of the 1976 Resource Conservation Recovery Act (RCRA), the 1976 Toxic Substances Control Act (TSCA), the 1980 Comprehensive Environmental Response Compensation and Liability Act (CERCLA or "Superfund"), and the 1990 Pollution Prevention Act (PPA).

The speaker will discuss these developments from her experience as an environmental attorney, but also, from her experience as a former chemical engineer and EPA Superfund On-Scene Coordinator with work history at several well-known NPL sites and also as a hazardous waste combustion permit writer for both EPA and industry.

381a Process Control Education: Past, Present, and Future

Thomas F. Edgar, Chemical Engineering, Dept. of Chem. Eng.,The University of Texas at Austin, 1 University Station Stop C0400, Austin, TX 78712

The practice of process control has undergone significant changes during the past 40 years, and with it the content of typical courses in process control has also evolved, albeit more slowly. In this paper, a critical examination of the current state of teaching undergraduate and graduate students in process control is provided. In the undergraduate course suggestions are offered for streamlining the presentation of certain topics in order to incorporate new material (e.g., biological engineering). The increasing role of simulation to validate theory is discussed. While there is a plethora of established textbooks for the undergraduate course, the situation for graduate level instruction has fewer options. The development of a virtual control book as an on-line repository of material contributed by a number of faculty around the world is one way to deal with the shortage of appropriate textbooks, as well as keeping such materials up to date. Undoubtedly future advances in techolology in control and instrumentation will occur, but the undergraduate course may not change so much, being restricted to its current three credit hour format.

Web Page: www.engr.utexas.edu/che/directories/faculty/edgar.cfm

381c Predictive Control from Practice to Theory

James B. Rawlings, University of Wisconsin-Madison, 1415 Engineering Drive, Madison, WI 53706

This presentation will describe the development of model predictive control in industry and academia since its start at Shell in the late 1970s.

381d Predicting the Future of Model Predictive Control

Manfred Morari, Automatic Control Laboratory, ETH Zurich, Zurich, Switzerland

This presentation will describe recent advances in model predictive control and postulate on future developments.

Web Page: control.ee.ethz.ch/~morari/

381e On Predicting the Future from the Past in Process Control

George Stephanopoulos, Massachusetts Institute of Technology, 77 Massachusetts Avenue, Room 66-440, Cambridge, MA 02139

The speaker will comment on the ability of a person to predict the future of process control based on the past. For example, would an engineer in the early 1970s have predicted what process control would look like today in industry or academia? Then emerging areas in systems and process control will be discussed.

Web Page: web.mit.edu/cheme/people/faculty/geosteph.html

381f Panel Discussion on the Past, Present, and Future of Process Control

Thomas F. Edgar, Chemical Engineering, Dept. of Chem. Eng.,The University of Texas at Austin, 1 University Station Stop C0400, Austin, TX 78712, James B. Rawlings, University of Wisconsin-Madison, 1415 Engineering Drive, Madison, WI 53706, **George Stephanopoulos**, Massachusetts Institute of Technology, 77 Massachusetts Avenue, Room 66-440, Cambridge, MA 02139, and Manfred Morari, Automatic Control Laboratory, ETH Zurich, Zurich, Switzerland.

This is a panel discussion of the past, present, and future of process control. The panelists will be the speakers in the session.

383a Progress and Outlook for Molecular Dynamics Simulation

Edward J. Maginn, Department of Chemical and Biomolecular Engineering, University of Notre Dame, 182 Fitzpatrick Hall, Notre Dame, IN 46556

Atomistic-level simulation of the condensed phase was one of the earliest applications of the digital computer. Alder and Wainwright at Livermore, Rahman at Argonne and Verlet at Yeshiva University were among the pioneers in the 1950s and 1960s in the use of a deterministic method called "molecular dynamics" to compute the properties of simple liquids. Most of these early studies were concerned with computing the properties of either idealized "hard" fluids or more realistic but still simple liquids such as argon. A major objective of this work was to compare the results of these computer "experiments" with predictions made from emergent liquid state theories. Since that time, advances in algorithms and techniques, coupled with an explosive growth in computer power, have led to the widespread use of atomistic simulations in the study of condensed phases. It is now common for molecular dynamics to be applied to complex systems such as proteins, surfactants, organic mixtures, electrolytes and polymers. While memory and processor speed limited the early simulations to a few hundred atoms, it is now commonplace to see simulations containing millions of atoms. In fact, using 64,000 processors on the new Blue Gene / Light computer, researchers at the Lawrence Livermore National Laboratory have carried out a simulation of 4 X 10^10 atoms! As important, the sophistication with which intermolecular interactions can be modelled has improved greatly, and the timescales over which molecular motions can be studied has increased substantially. In this talk, we will review the history of molecular dynamics, highlight the role chemists and chemical engineers have played in developing molecular dynamics into a useful tool, and try to predict what some of the applications of molecular dynamics will be in the future.

383b Molecular Simulation Methods for Transport Properties and Processes

Peter T. Cummings, Department of Chemical Engineering, Vanderbilt University, Nashville, TN 37235-1604

Text Not Available.

383c Progress and Outlook for Mesoscale Simulation

Juan J. De Pablo, Department of Chemical and Biological Engineering, University of Wisconsin, 1415 Engineering Drive, Madison, WI 53706

Text Not Available.

383d Commercial Tools for Molecular Simulation: Current, Intermediate and Long Term Perspectives

David Rigby, Accelrys Inc, 10188 Telesis Ct, San Diego, CA 92121

Text Not Available.

477a Fluid Mechanics and Transport Processes: From Colloid Science to Small Devices

Howard Stone, School of Engineering and Applied Sciences, Harvard University, 40 Oxford St., Rm 224, Cambridge, MA 02138

We discuss recent themes in fluid mechanics of small-scale systems, ranging from multiphase flows in small devices to wetting of topographically and chemically patterned substrates. The ideas span basic fluid dynamics of confined systems to new applications inspired by the design of new kinds of materials.

477b Oriented Assembly of Anisotropic Particles by Capillary Interactions

Kathleen J. Stebe[1], Eric Lewandowski[1], Jorge A. Bernate[2], and Peter C. Searson[3]. (1) Chemical and Biomolecular Engineering, Johns Hopkins University, 3400 North Charles Street, Rm#221, Maryland Hall, Chembe, Baltimore, MD 21218, (2) Chemical and Biomolecular Engineering Department, Johns Hopkins University, 221 Maryland Hall 3400 North Charles Street, Baltimore, MD 21218, (3) Johns Hopkins University, 3400 N. Charles St, Baltimore, MD 21218

Recent advances in micro- and nanoscale science exploit self-assembly to create ordered structures from micro- and nano-scale objects. The ability to dictate the orientation of an object, or to force an object to assemble in registry with other particles remain important goals in the field. We study capillary interactions that drive oriented assembly of anisotropically shaped objects. Capillary interactions between particles drive particle assembly at fluid interfaces. The particles distort the interface, creating excess surface area. Excess area decreases as particles approach each other, creating attraction. Anisotropically shaped particles create excess areas that are non-uniformly distributed around the particle's center of mass. Local regions of concentrated excess area develop that drive oriented assembly. We use a force balance to develop relationships between particle aspect ratio and regions of concentrated excess area that drive orientation and assembly. These arguments are supported in experiment. Short-range capillary interactions have been predicted to promote alignment of particle faces. Based on this concept, particles with complex end faces that align with end-to-end registry are created.

477c Randomness and Runniness of Confined Fluids

Thomas M. Truskett, Department of Chemical Engineering and Institute for Theoretical Chemistry, The University of Texas at Austin, 1 University Station, C0400, Austin, TX 78712

Fluids trapped in small spaces feature prominently in science and technology, and understanding their properties is critical to progress in fields that range from cell biology to the engineering of nanoscale devices. Such "confined" fluids behave differently than bulk samples. Properties strongly affected by confinement include (i) how the particles pack together, (ii) how the thermodynamic state of the fluid is modified by squeezing or heating, and (iii) how (or how rapidly) the fluid mixes or flows. The first two have been reasonably well understood for some time, but a general rule for even qualitatively predicting the third has proven elusive. In this talk, we highlight how recent theory and molecular simulation have discovered quantitative links between the entropy ("randomness") and the dynamics ("runniness") of confined pure fluids and their mixtures. These links provide new physical insights and allow one to readily predict how confinement will impact the transport coefficients of bulk fluids.

477d Using Quantitative Modeling of Molecular Transport to Accelerate Development of Membranes for Gas Separations

David S. Sholl, School of Chemical and Biomolecular Engineering, Georgia Institute of Technology, Atlanta, GA 30332

Membrane-based separations have the potential to make enormous contributions in energy-related applications involving, for example, carbon dioxide capture and hydrogen purification. The fabrication and testing of non-polymeric membranes from new materials is extremely time consuming, so the risk associated with the experimental development of a genuinely new material is high. If quantitative models for the performance of new membrane materials can be developed in the absence of direct experimental data, these models can play an important role in accelerating practical development of the most promising materials. Several examples where atomically-detailed simulations have been used for this purpose will be discussed, including dense amorphous metal films for hydrogen purification and metal-organic framework materials for carbon dioxide separations.

AUTHOR INDEX

AUTHOR INDEX

AUTHOR INDEX

American Institute of Chemical Engineers
3 Park Avenue
New York, New York 10016-5991

proceedings.com